MANAGING IT PROJECTS

HOW TO PRAGMATICALLY DELIVER PROJECTS FOR EXTERNAL CUSTOMERS

Marcin Dąbrowski

Apress®

Managing IT Projects: How to Pragmatically Deliver Projects for External Customers

Marcin Dąbrowski
Nowy Wiśnicz, Poland

ISBN-13 (pbk): 978-1-4842-9242-6 ISBN-13 (electronic): 978-1-4842-9243-3
https://doi.org/10.1007/978-1-4842-9243-3

Managing Director, Apress Media LLC: Welmoed Spahr
Acquisitions Editor: Shiva Ramachandran
Development Editor: Jim Markham
Editorial Assistant: Gryffin Winkler

Cover image designed by Isaac Soler at eStudioCalamar

Distributed to the book trade worldwide by Springer Science+Business Media New York, 233 Spring Street, 6th Floor, New York, NY 10013. Phone 1-800-SPRINGER, fax (201) 348-4505, e-mail orders-ny@springer-sbm.com, or visit www.springeronline.com. Apress Media, LLC is a California LLC and the sole member (owner) is Springer Science + Business Media Finance Inc (SSBM Finance Inc). SSBM Finance Inc is a **Delaware** corporation.

For information on translations, please e-mail booktranslations@springernature.com; for reprint, paperback, or audio rights, please e-mail bookpermissions@springernature.com.

Apress titles may be purchased in bulk for academic, corporate, or promotional use. eBook versions and licenses are also available for most titles. For more information, reference our Print and eBook Bulk Sales web page at http://www.apress.com/bulk-sales.

Any source code or other supplementary material referenced by the author in this book is available to readers on GitHub.

Printed on acid-free paper

I dedicate this book to my wife – for her patience, understanding, support, and warmth in tough times.

Contents

About the Author

Marcin Dąbrowski managed and supervised large IT projects for banks and the biggest telecommunications groups in the world (including T-Mobile, Orange, Telefonica, O2, Vodafone, Hutchinson) in various parts of the world (Europe, South America, Middle East, Asia, and the Far East). He cooperated with clients representing extreme national and organizational cultures. Marcin was responsible for all parts of the IT organizations (up to 1,400 people), from sales, consulting, and marketing, through software development and production departments, to implementation and maintenance. He started his career as a software developer and later worked in practically all roles of the IT organizations, ending as a VP and board member of two publicly listed software providers

In the years 2007–2010, he conducted research at the Department of Telecommunications at AGH University of Science and Technology. At that time, he published scientific articles, among others, for ACM CCS DIM 2008, IEEE ICCS 2008, SECURWARE 2008, and China Communications. He was also the author of contributions to ITU-T.

Marcin Dąbrowski is a graduate of the IESE Business School and the AGH University of Science and Technology. He holds an MSc in Electronics and Telecommunication and an MSc in Computer Science.

Endorsements

An extremely powerful book which describes well the highlights of critical project management well supported by real-life experience.

I read it with the intent to learn and the curiosity to see if I could find my 22 years of experience and my learning about managing large technological transformations, and I have to say that I could relate to many of the experiences and examples that have been brought forward.

I find the book very well written with a mixture of technical teachings, advice, tips, and sharing profound experience always supported by the real cases to keep the learning imprinted in the mind of the reader, proving a sense of life and not of abstract theory.

Reading this book would be beneficial for many players involved in major technological transformation given that it does not only cover all technological phases of projects but also the areas of communication, leadership, resource management, short-term vs. long-term goals and the knowledge trade-offs vs. investing in talent.

A piece of experience lived on the battlefield described with simplicity and clarity; very inspirational.

I love the style of communicating; very complex concepts are described in a light and easy-to-understand way, which makes the reading enjoyable and efficient at the same time.

I completely agree with the consideration that focusing solely on achieving sales objectives at any price leads directly to uneconomic and delayed projects, and I love the focus on the involvement of all parties of the organization in the initial phase, especially the departments responsible for implementation.

The recommendation not to make savings on the costs of the domain experts either during the sales process or during the project is a rule to follow, quite often ignored nowadays in many transformations and project executions. And the consequences are very well described: "The losses will definitely exceed the savings."

My favorite chapter emphasizes attitude, customer orientation, willingness to solve stakeholder problems as a major source of success. In addition to investing in an on-site team, this strengthens relationships with customers. You got it right!

I was astonished at the clarity of the expression of some of the most common mistakes of those days as dealing with agile methods as the equivalent of acting without scope, time, and schedule, the conflict between the flexible scope and the fixed price or the failure due to lack of alignment between the chosen project management methodology and organizational culture.

What can I tell you? I think this session provides many answers to many of the struggles that many businesses and industries face.

Managing IT Projects is a key book; it contains a comprehensive overview of the do's and don'ts, tips and tricks that can be applied to each technological transformation addressing not only the technical steps but considering people, culture, and communication.

A readymade guide to success.

Antonietta Mastroianni

Chief Digital & IT Officer at Proximus,
Telco Woman of the year 2022

Straight to the point, well written, and bold!

A complete and holistic guide to managing complex transformation projects with customers with a lot of real-life examples.

The book shows how to successfully deliver projects, what to focus on, and what to avoid. It goes far beyond abstract project management theory.

One of the best books on project management!

Dr. Joerg Storm

Global Head IT Infrastructure Mercedes-Benz Mobility

This is a book of incredible value for experienced project managers, stakeholders, business unit leaders, CEOs, and junior professionals starting in their careers. Many books have been written on Project Management in general and even a standard Body of Knowledge (PMBOK) has been created to help certify the profession. As a certified Project Management Professional and having over 20 years of experience in the IT industry, I have never seen any comparable summary of root causes for cost overruns, delay typology, and associated quality issues, including potential relevant remediation/avoidance measures stated in such a clear and pragmatic approach.

Moreover, the author includes a no-nonsense, reality-based view of the mindsets, dysfunctional incentives, and conflicting priorities in the IT project business, giving the reader an incredible level of insights from both vantage points at the same time: the customer and vendor points of view. All of it narrated in an electrifying first-person perspective along with experience and an abundance of details, making it overall a great reading.

Absolutely recommended book for all professionals in the IT industry!

Francis Cepero

IoT Business Unit Leader A1 Group,
former VP Innovation & Technology at SAP

IT projects are very often completed with big delays and exceed the projected budget by far! These situations of frustration, distrust, and last but not least the resulting economic disasters for customers and suppliers are well known. Do we have to accept that as a fact and simply take it into account in the planning? Is the IT landscape unreliable in general, are agile development methods unpredictable, or is the project management just incompetent?

Marcin Dąbrowski gets down to the essence of the matter and works out step by step the mechanisms and forces acting between the cooperating parties. He analyzes the impact chains in all project phases and uses concrete examples to compare the desired and the real results. He opens the reader's eyes to the many hidden traps in one's own way of thinking, regardless of one's role in the project.

The book is impartial and ruthlessly honest, with the focus on the step toward an overall higher added value. It suggests in a comprehensible way how conflicting interests can be aligned toward a common goal and how efficiency can be significantly increased in this way. The author's extensive experience in the implementation of IT projects is reflected in the applicability and practical orientation of the solution approaches.

Managing IT Projects by Marcin Dąbrowski is a reference work, a textbook for project management in general and for the implementation of IT projects in particular. It states clearly that the challenges can be mastered by management expertise and shows how.

The best book I read this year.

Holger Weichhaus

Senior Director for Development Quality, AUDI AG

In this book, Marcin Dąbrowski tackles the large gap between the theory and practice of Project Management. Based on his many years of experience, the author identifies the various problems and conflict situations that can have huge impacts on the projects, and presents practical and field-proven solutions and frameworks for those problems.

Reading it now, refreshed many concepts that I had forgotten, and shed light on many problems that I had identified along the years. This is a very practical and insightful book that I wish I could have read when I started working in project management.

I will definitely make it mandatory reading for all my project managers.

Ricardo Campos

CEO ITSCredit, former CEO at ActivoBank and
Chief Digital Officer at Millennium Bank

Having been on the receiving end of many IT-projects that passed my desk and being involved during escalations with vendors or internal teams, I have seen the hard way that only very few people truly understand the complexity of managing projects and stakeholders. The phrase "double the cost and add 20%" as a reflection of Parkinson's law I have seen many times when it came to calculate the true cost of a project in hindsight. Marcin's book really is addressing in a very practical sense what the reasons for many of the disappointments on both the customer and the supplier are.

I wish I would have been able to profit from his experience maybe 20 years earlier. I would recommend it for anyone that is part of the management team or project management of an IT-supplier or customers dealing with large scale IT-innovation.

The book really draws you into how to successfully manage an IT-project both on a very practical note and also how to influence both internal and external stakeholders.

Brian Stout

Former CEO of BT One Phone, board member
of T-Mobile The Netherlands, and CEO of KPN Belgium

There are many books trying to explain to us how to design, sell, and deliver transformative IT programs for demanding clients. *Managing IT Projects* is more. Marcin has created a masterpiece for all the IT professionals who want to learn from other people's errors and do it right in the first intent. He excels in putting the attitude first. After reading his book in one night, I will never forget the approach it left in my brain: "Prepare for the client, connect deeply, build relationships and play." Marcin takes you to all the stakeholders in the program life cycle and helps you to understand how they think and act. And he teaches us how to influence their thinking and acting positively to create a joint success story.

Outstanding value for practitioners.

Michael Brandenburg

GM 2Quadrate GmbH, Lead Coach at IESE Business School,
former VP at Capgemini

Are you looking for a book that explains the journey behind an IT project and how one deals with the challenges that come along the way? Look no further – Marcin captures this brilliantly in his book *Managing IT Projects*. He shares his experience and valuable lessons from different situations that have arisen over the years in the relationship between supplier and buyer. He gives the reader great tips to shorten that learning curve and avoid some common mistakes along the way. In my experience, all projects should be approached as business projects, although they rely on IT resources. Marcin touches on this and his valuable sharing focuses on the human aspect, which are values that will never go away despite technological advancement.

Congratulations on this great book Marcin and to you the reader – I encourage you to take the shortcut from mistakes and bad decision-making and learn from his experience.

Gudny Helga Herbertsdottir

CCO of VIS Insurances

Experienced manager and leader with demonstrated results in Digitalization, Strategy, Results Delivery, and Marketing. Implementation of Digital VÍS, a methodology used for the digital transformation of the largest insurance company in Iceland with over a 100-year-old history.

Describing this book as the ultimate practical manual for effective IT project management, or as a fuss-free guide to dealing with the grievances of every project is falling short. Marcin spits out, straight-to-the-point, years of remarkable project management experience in a format that should become the bedside book of every project manager, but also for every person in an IT organization.

When drilling down the details of each pathology, exposing its causes, and providing real-world advice for solutions, the author made me want to have every actor involved in the process read it as it becomes clear not even the most experienced project manager can handle it all.

While customers still believe they can outsmart their providers to get good, cheap, and quick in every project, Marcin gives undeniable examples that demystify the unicorn and show that unless the objectives are really aligned, no success is possible.

To share all this experience and knowledge is a very generous gift by the author and I recommend anyone starting their career to devour this book to get an invaluable head start and every experienced professional to delight themselves with relatable anecdotes and certainly more than a handful recipes of how to tackle challenges that you either already faced in your journey or most definitely will encounter in your next project.

Fabio de Souza

Head of Solution Architecture at Amdocs

As a starting project manager in the year 2000, I wish I would have had this insight in my hands. I would have avoided a lot of bizarre and difficult meetings from the beginning.

But as far as I see it, this is not just for project managers but also for middle management. In the end, short-term bliss becomes an obligation to deliver. And for top management, I have been both on the buyer and seller side. We all need to optimize the EBITDA or EBIT, but we need to consider why we engage in the project in the first place. We want results and that is why there is a business case. The real benefit comes from the end result. Whether we get it 10% cheaper with, in reality, 30% more cost with a delay, or we over-promise as a supplier and the end result is the same, none are happy – especially, neither the shareholders nor the employees. I know, it is tricky in the competitive landscape, but that is the reality. You can build a freeway over the cemetery once, but try that another time.

Jarno Kekäläinen

CEO at Telia Inmics-Nebula

A thorough analysis of the IT procurement to delivery life cycle written from the perspective of the supplier. The book is full of hints and tips for how to succeed and avoid the many pitfalls that await the unwary. The book is recommended reading for IT professionals on either the client or supplier side, especially those with only a handful of projects under their belt.

Mike Corrigall

Former Vice President of T-Mobile International

A must read for project managers who want to overcome critical situations in the life cycle of a project/project implementation and make it happen in time and in budget. Marcin is sharing with us his insights of many years of experience, not concentrating on the theories and tools we all know but on real life and how to overcome the inevitable hurdles appearing in every complex project implementation.

Take the time to read this book; it will save you from many pitfalls along the way!

Patric Thate

CFO UBM Development AG

The costs of delayed or failed projects can far outweigh the initial project cost so readers would do well to take heed. Marcin Dabrowski demonstrates a wealth of practical experience in his book as he documents the many challenges, potential planning failures, and frustrations faced by IT project managers. He uses his experience to supplement theoretical project management with real-life examples and guidance for project teams.

The book lays out the many stages of a project life cycle, especially the early stages of sales and project setup, and in particular addressing some of the early pitfalls that can cause significant problems further down the line.

The step-by-step approach to the life cycle makes this an ideal support both to new project managers and for the more experienced.

Oliver Matthews

Vice-President & CMO, Frankfurt School of Finance & Management

With honesty and candor, Marcin offers his vision of how to navigate in the turbulent waters of managing large and complex projects based on his long experience and always from the management point of view. This book does not offer a sweetened and theoretical vision of this type of project and of the situations (many times unwanted) that the team that manages these projects will encounter. Nor is it intended to recommend a management model or offer magic recipes (in my experience, I have never seen a solution that works for all situations and customers). What the reader will find in this book are examples of real situations that Marcin has experienced in different projects and a reflection on what decisions have led to some of these situations. In some cases, the final conclusion was satisfactory as the project team managed to bring the project to success, but in others, the projects ended in failure. Managing a complex IT project is not an exact science, although knowing and applying good practices and types of management that have worked in other cases is always a great help. In order to successfully manage this type of project, it is essential to have a team of people with extensive knowledge and experience both in the field in which the project is to be developed and in managing customer relationships. Like any project, these projects are going to be used by people and are intended to solve business problems for a client. The perception of what customers and end users will have of the project and its final result will be influenced by subjective criteria and by the attitude and aptitude that the project team demonstrates. The best recommendation to face a complex project is to do it using common sense, empathy, investing in a good team from the beginning, and being aware that, throughout the life cycle of the project, we will find difficult situations that an experienced team will have more options to resolve successfully.

So, either if you are starting out in project management or just want to read about what can happen in real complex projects, you will be able to benefit from the experiences shared by Marcin in this book and... good luck!

Laura María Alcón del Lago

Former Project Manager and Solution Architect at Telefónica

I traveled the world and delivered projects of different shapes and sizes, in the telecom industry for years. From Iceland to Afghanistan, from Norway to South Africa. Believe it or not, but all these projects suffered from similar issues. And...they were not technology. The role of sales and the sales process that is often limited to the handover to the project manager (in the form of a project chart) is a huge underrating of the criticality of this phase. What really matters on a project – is it a methodology, the technology, the tools, the carefully selected team, or communication skills? How does it affect the project's chances to be successful?

I can't remember another book in the business area that took me through such a wide range of emotions and brought back so many memories!

The strength of this book is its matter-of-factness – it focuses on what really makes a difference. Good for the junior project managers (to avoid some bruises), excellent for the seasoned ones (to find a permanent recipe).

Carefully selected pieces of information, written by a seasoned project manager to help other project managers grow and mature.

I wish I wrote this book myself!

Rafał Gruszczyński

Site Manager/Project Manager at Spyrosoft

The author has shared with the readers his extensive experience in strategic project management that he gained while working for international organizations. He surprises with his ability to poignantly synthesize the assessment of complex situations and his accurate analysis of their causes which include the complexity of the parties involved in the process, the psychology of its main actors, sometimes conflicting goals in organizations as well as differently understood assumptions and rules seemingly obvious for everybody.

The book contains plenty of case studies from the author's extensive experience, which makes it very interesting and also enables the reader to identify with a decision maker managing large projects, changing the reader's perspective, his perception, and thinking about a lot of things. Various situations and different ways in which a project may evolve as well as escalating problems presented in the book make the reader think about the mechanisms and consequences of the decisions that are made. The topics discussed in the book are incredibly pragmatic and versatile, they touch upon all aspects of a project – from its origins till its completion. The book is not a theoretical reference material describing different methodologies, it is based on business psychology, emotions, and conflicting interests – and all of this makes it a worthwhile read.

After reading the book, you'll definitely ask yourself: Why are we still making so many mistakes if the behaviors and their consequences are so obvious and predictable?

Piotr Piątosa

CEO Ant Solutions, former CEO Comarch Healthcare

A very interesting and much-needed book for everybody who wants to skillfully manage or consciously participate in projects. The author's precise analyses are based on his extensive experience.

The coherent synthesis presented in the form of clear suggestions enables the reader to avoid mistakes, but first of all fully understand what happens in the project at its different stages, and aptly manage difficult situations.

The presented methodology overlaps with my experience as a person who supervised projects on the client's side – beginning from the specialist level to a board member. The approach that I applied in many projects guided by my intuition is explained in this book in a rational and convincing way.

At the same time, this book is a perfect tool for people who "switch sides" transforming from a client to a supplier, it provides plenty of practical solutions and tears down the "towers of fear."

It is an obligatory read for all project parties and in any region of the world. Due to its comprehensive approach, the book is a suitable read for people having some project experience. For beginners, it can be a manual of sorts for their first steps in projects.

It is an easy read, the argumentation is clear, supported by examples and summaries of conclusions. It's a book that can be and should be read multiple times. It is also helpful for people like me who have left their corporate jobs on the client's side and became advisors, consultants, and freelancers on the supplier's side. Seasoned project participants will find here a lot of observations they will agree with, but the value of this book lies in its comprehensive nature and in the correlation of a number of issues.

I encourage you to read this book. It's one of the best business books that I've had the pleasure to read. I recommend it not only in relation to IT projects but also to other projects in the times of advancing digitization.

Rafał Markiewicz

Former CTO E-Plus Gruppe & KPN International,
CTO BASE, CEO Nexus Telecom

I've read the new book *Managing IT Projects* by Marcin Dąbrowski on IT project management with great interest. Unlike many other books available on the market, it doesn't talk about project methodologies, but focuses primarily on the practical aspects of the supplier–client cooperation.

Marcin presents the project life cycle from the supplier's perspective, he thoroughly analyses each phase from the pricing stage, sales, through the implementation phase to the final success or possibly delays, penalties, or a contract breach. Each of the phases is analyzed in detail, which shows precisely what happens both on the client's and supplier's side.

I recommend this book especially to all managers working in companies commissioning IT projects. It will help them to better understand their suppliers, their modes of operation, problems, and consequently select more effective management methods for projects carried out by external suppliers. After reading this book, a lot of them may discover why the practices used up to now have not produced the expected results.

It is a book worth promoting as there's nothing similar on the market yet.

Marek Lenz

CIO Allianz Poland

Everlasting delay. Almost everyone who has managed projects has experienced that. This book can help you learn about the reasons for the delays as well as the ways to eliminate them on the supplier's side.

It is quite conspicuous that most emphasis is put on the supplier's perspective, yet it's not a limitation – it's a perfect resource for a client who wants to understand the mechanisms of issue generation and thus work on some ways which will allow them to prevent delays. That's why it's a material for analysis not only by suppliers of IT solutions, but also or maybe even primarily for their clients. I can imagine a team of managers who – after reading this book and learning about the bleak and rarely discussed reality of the projects – will be able to run a project in such a way that they can avoid the trap of delays.

A word of warning! The read may get a little frustrating as you'll be reading about the things that you often see in your work on a daily basis.

Bartłomiej Irzyński

CTO Vectra | Multimedia Polska

How many projects end up with dissatisfaction, exceeded budget, and underimplemented scope...

The book is primarily for decision makers and project managers of solution and integration services providers in large IT projects. It is a kind of prescription inventory for preventing key risks in preparing and carrying out projects. It's also a valuable read for customers who can learn and understand the perspective of IT solution providers.

Recommend reading.

Dorota Poniatowska-Mańczak

CIO Credit Agricole Bank Poland

Marcin skillfully describes the reality of running big and complex IT projects. He perfectly identifies the basic, most common mistakes and shows how to avoid them.

I wish I could have read this book 20 years ago. The book, quite like the aviation law, is written in suppliers' blood. The author accurately claims that an ethical and wise purchasing process determines the success or failure of the project from the outset. Thus, most attention should be paid to commissioning and contracting.

Fortunately, modern businesses lean more and more toward agile project management, which means that the process of managing change and client's expectations is close to reality. Agile works wonders in the relation between the client and the supplier.

Rafał Styczeń

Founder & CEO, Ailleron | IIF

While reading this book, in my mind, I constantly stumbled upon the word "honesty." It is present in virtually every issue discussed. Honesty first – toward the client, but also toward one another within the company. For me, honesty and mutual respect are very important values. I like the fact that Marcin promotes them in his book.

It's also a great read, it's lightweight – it's written in a simple language, with no pompousness or jargon.

The author paints a complete picture, in which he depicts in a skillful and professional manner both the well-known processes, such as project phases, and the unwritten rules that often decide about success or failure.

While reading the book, it crossed my mind multiple times that the engineers dealing with clients should read it. It's incredibly practical. It wins my trust.

Ewa Maciaś

Director of Engineering at Nomagic,
ex-Director of Engineering at Google

When I started reading *Managing IT Projects*, I was wondering how it might be different from other similar books on the market. And I was positively surprised. From the very first pages, it becomes obvious that the book is written by a person with many years of experience in running big projects, and also someone who likes sharing this experience. The author doesn't focus on project management theory, he shows the importance of a holistic approach to all aspects of the project, from the beginning of the sales process to the project's end.

The book clearly shows what should be done to bring a project to a successful end but also what to avoid. It's also worth mentioning that the great advantage of this book is that it's pleasant to read.

Leszek Cygan

Senior Manager, Kyndryl (formerly IBM)

A very interesting book comprehensively describing the process of executing IT projects. It also talks about phenomena and areas which could be described as a "grey zone," meaning, for example, the gap between goals and actions in a company, the dynamics of the supplier–client relations, overlapping competencies within the company, etc.

There are not many books which deal with these areas so openly and that's why I think that author's observations are worthwhile.

Lucjan Sosna

Senior Director Software Engineering,
CAE Flight Services Poland

Effective and straight to the target!

The book you're holding was written to fill the gap between theory and practice. It's not another handbook. It's a guidebook you can use anytime to find the right solution and inspiration.

You'll find here practical and effective tools from the business and organizational areas, but also advice on people skills – supported by multiple examples from the author's everyday life.

The author is an experienced manager who carried out many international projects. He draws upon his knowledge and best practices, but also on his own mistakes. The power of this book is that it makes you pause and reconsider the way you act.

It's a must for People Leader!

Magdalena Firek-Gundelach,

Head of People Ailleron, Global HR Manager
Brown Brothers Harriman, HR Manager Philip Morris Poland

Each project we embark on is a journey. It has a clear starting point, we reach certain places on the way that could be described as milestones, which in turn allow us to reach the finish line. Sometimes the journey can be a proverbial walk in the park when we know exactly how to get there, what we are going to encounter on the way, all is good, and the sun is shining. However, more often than not, the project is taking us to unchartered territories akin to the alpine trail. No matter how well we think we are prepared, how much effort we put in planning the route, there will be surprises: be it sudden change of the weather or the trail we expected to follow is closed or maybe there is a patch of snow and unfortunately, we have forgotten to pack an ice axe and our crampons are not fully functioning. How the organization is facing challenges, how flexible it is in finding solutions to the problem at hand will make a difference between getting there (success) or failure of the given endeavor. No publication or training course will prepare us for what lies ahead, but learning from experience of project veterans is like getting a critical piece of equipment we can use over and over again.

Managing IT Projects is not another book about the project management methodology. It is a collection of well written observations, supported by real-life examples. While the author is clearly focused on the IT domain, I was very surprised how relevant the specific topics are to projects in financial servicing or for that matter to any industry. Of course, there are major differences related to things like payment schedules, penalties, re-negotiation of contracts, etc., but the advice given is simply relevant and on point. While reading, it was fascinating to find yet another "aha moment" – I've been there in my project career before, I have faced similar challenges, why have I not thought about it before?

Whether you are a seasoned project management professional or just learning the ropes, *Managing IT Projects. How to Pragmatically Deliver Projects for External Customers* should make it on your recommended reading list.

Jaroslaw Jasinski

VP, Implementation lead, Financial industry – front-, middle-, and back-office program management

Foreword

When I started learning about information systems in the 1980s, it became clear that there was a significant discrepancy between the theoretical world of systems development and daily reality. At that time, many may have thought that the field of information systems was still young and needed a few more years to mature, closing the gap between planning and reality.

Four decades later, this gap between planning and reality still exists, leading to frequent disappointment and suboptimal use of resources. In 2020, BCG estimated that, on average, 70% of IT project implementations are not successful. Given that the global IT outsourcing market is larger than $700 billion, the size of lost resources due to unsuccessful technology projects is staggering. If an example is needed, I recently heard a story about the replacement of a legacy system at a large airline company that began ten years ago; the current expectation is to finish the project in the next six months…

It is likely that disappointing implementation projects will continue to be part of our reality for the foreseeable future. However, this book, *Managing IT Projects: How to Pragmatically Deliver Projects for External Customers*, offers practical advice based on 20 years of experience to help IT managers overcome the problems that will inevitably arise in IT projects.

Written for IT project managers who manage projects for external customers, this book emphasizes the importance of the sales process and project initiation for eventual project success. With practical examples, it covers all the necessary steps to increase the likelihood of success. The author, a vice-president of an IT company with 20 years of experience in the technology sector, emphasizes the importance of the social and relational aspects of interacting with clients. The book helps readers understand how to communicate more effectively with clients to set realistic expectations and how to navigate unexpected situations together, rather than as opponents.

As the central role of information systems continues to grow in society and business, it is essential that we continue to learn how to benefit from IT systems without being hindered by implementation problems. The only way to make progress in this area is by continuing to improve our theories and models of systems development and by complementing these theories and models with practical guidance and advice. This book provides valuable insights in this regard, helping not only individual IT managers to perform better but also allowing IT companies to build a competitive advantage and ultimately enabling society to benefit from the fruits of information systems.

Prof. Govert Vroom

Strategic Management Department
IESE Business School

Introduction

Constant project delays are without question one of the biggest problems in the business world. Yet, there are almost no books addressing the real root causes of this problem.

Most project management books focus on PM methodologies which tell you how to work but are not even close to explaining the real mechanics of project management with complexities of customer–provider environment. Therefore people implement novel project management techniques and projects are still delayed. They do not understand what is important for customers. They deliver what is specified in the contract, but customers are still not satisfied and break contracts. Customers threaten with penalties or sue providers and the methodologies give no answer what to do in such situations.

Most providers and customers are afraid to mention their real behaviors or unfair games – they avoid admitting to practices that later are the real causes of project disasters. This book not only addresses these problems point blank but also gives pragmatic answers on how to act and be successful in real-life situations.

This special compendium of knowledge, "project management black magic," contains proven recipes for project managers and IT organization managers, as well as advice on how to act in critical situations and how to avoid

- Underestimating projects
- Delays in their implementation
- Disputes between customer and the project team
- Conflicts with other project managers, implementation departments, and the sales team
- Misunderstandings in customer relations
- Delivering underdeveloped products
- All kinds of threats, from penalties for delays to breaking the contract

The book is a complete manual for all critical situations along the project's life cycle. Each chapter contains real-life examples with analysis and guidelines for the reader. The book contains down-to-earth knowledge that usually project managers learn over the course of years in critical and stressful situations with no preparation beforehand.

This book is not about academic terminology, processes, and tools – it's about pragmatic strategy and philosophy for successful project delivery and cooperation with customers.

By the end of this book, the reader will understand

- What are the main reasons of the delays in projects, where do they come from, and why it has nothing to do with what is usually openly written in most of the books
- What are the key success factors in managing projects and why it has nothing to do with project management methodologies and techniques
- Which are the critical moments of the project in real life and how to prepare for them
- What are the usual expectations of customers and why it has nothing to do with project results
- How volatile the project goals are and why the only thing that matters is customer understanding and proactive attitude
- How customers can threaten with penalties to get what they want
- How and when customers can cancel projects or sue the provider even if everything is being delivered on schedule

And the reader will be able to

- Understand the whole project life cycle from a holistic point of view (from sales process to project closure)
- Achieve customer's satisfaction and support independent of a project's results (even during project crisis and facing huge delays)
- Transform the customer into your best friend and project supporter

Effectively communicate with customers to make things happen

Understand when and how to communicate "bad news"

Identify all potential moments of crisis in a project's life (internal and in cooperation with customers)

Use the tools and techniques necessary to always be prepared for the worst-case scenarios

Fight their way out of huge project delays

Defend against penalties and court proceedings

Understand when continuation of a project is pointless and when the project should be stopped

How and Why This Book Was Written

Since the beginning of my career in IT, working in different roles, I've seen the same, recurring problems. While changing positions, I was slowly gathering the pieces of the puzzle, bit by bit. Every time gaining new experience cost me overtime, nerve-racking situations, fatigue, stress.

The worst thing was that nobody had prepared me for my struggles.

I remember the situation that completely changed my perception of project management. In one of the projects, we were preparing for the steering committee, together with the customer's team. The customer was pushing us to extend the project schedule by one year explaining that it is necessary for both sides and this is the only realistic approach to continue. Next day, during the first five minutes of the steering committee, in front of our CEO, the customer stopped the project and broke the contract! Official reason: expected delay of one year. I was lost. I didn't understand what happened and why it happened.

While the technology or management methods are described in multiple resources, it's difficult to find books about what you shouldn't tell your clients, how to explain delays, what to do when a client threatens to charge penalties, how to deal with a project which generates losses, or with a situation when the client forces you to work faster to make up for delays and your own management takes away the best people from the team and moves them to more profitable projects. In particular, nobody is willing to write about, or even explain, how to fight against all sorts of "pathologies" inside the organization, the behaviors of some individuals inside the company and on the client's side which effectively hinder progress in the long run, destroy teams, and lead to the best people suffering from burnout.

When I finally managed to understand most mechanisms causing problems and delays in projects, the reasons for friction in the cooperation with the client or between different departments within the organization, I more and more observed that managers of the projects I supervised had to go through the same ordeal that I had gone through. They also had to learn on the job, pay with their own "blood and sweat." While I was trying to explain and help them, I often thought how much easier it would be to describe my own experiences and share them in the way that would save them the hassle of learning everything from scratch and help them avoid making the same mistakes.

For years I'd been promising myself that I'd do that one day, that I'd transfer my thoughts to paper. Obviously, in the face of the current workload, all personal plans are usually put off until some indefinite time in the future. And finally there came a time when I decided to slow down, take a longer break. As it usually happens, the long break was very short. But while it lasted, I developed a habit of writing systematically to my morning coffee. And this is how this book was created in three months.

I realize that I only touched the tip of the iceberg, that the problems described here can be talked about endlessly. Nevertheless, I hope that it's a good starting point.

This book is based exclusively on my practical experience, on the mistakes I made, on the bitter pills of failure that I had to swallow, on the successes that I could celebrate, on the projects that I was happy to be a part of. It contains advice and lessons that I learnt from the people that I had the pleasure to work with or against whom I had to (professionally) fight.

I do hope that this book will let the reader avoid many mistakes, get some distance, and most of all understand that projects are run by "simply" human beings, that a good conversation, a little empathy and devotion to the client, and a proactive approach can work miracles, and that there are some deadlock situations that you just can't get out of and sadly you have to give up.

Marcin Dąbrowski, 3rd June 2021

CHAPTER

1

When Does a Project Really Start?

Anything that begins well ends badly. Anything that begins badly ends worse.

—Pudder's Law

According to most textbooks, a project starts when the contract is signed. After that there's a short period of celebrating the success and overarching excitement. Within a week or two, a project manager is appointed and the organization gets down to work. The right departments select the people, domain experts are identified. However, several weeks pass and things start getting out of control, project management theory proves inadequate. It's at this moment that the project manager "discovers" to his surprise that

- The project schedule is unrealistic – despite the fact that only a short time elapsed since the commitment, already months of delay need to be taken into account.

M. Dąbrowski, *Managing IT Projects*,
https://doi.org/10.1007/978-1-4842-9243-3_1

- Deliverables don't exist or a lot of people and a significant, unaccounted for amount of time will be needed to develop them.
- The scope of work is much wider than initially set out in the contract, some person made wrong assumptions while preparing the offer, but the person in question either no longer works for the company, is too busy, or unwilling to support the current project.
- Department heads who should take responsibility for delivering specific parts of the scope refuse to approve the project schedule, claiming that it's unrealistic and also that it had not been consulted with them during the sales process and thus they're unable to accept ridiculous requirements (scope, time). What's more, the project manager knows that sadly they're right.
- There're not enough experienced domain experts, so the manager finds it difficult to find adequately qualified people for his project and has a lot of problems with setting up the project team. Line managers don't want to put their lucrative, well-functioning businesses at risk. They don't want to let go of "their" best people for a new, unknown (and surely precarious) project. The clue is that they don't want to jeopardize their annual bonuses for some new enterprise which stays beyond their control and responsibility.
- The agreed price or the so-called anticipated income does not cover the anticipated project costs, including the investment that the company will have to make, especially at the initial stages of development. The project manager reaches the people who prepared the estimates and it turns out that they are not accountable for the performance of specific departments, and their main focus is meeting sales targets.

The preceding scenario is the gloomy reality. At first glance, it's clear that most of the problems are sort of "inherited" by the project after the sales process. **The complications to a large extent result from an inappropriately conducted sales process and the people who are later blamed are usually the project manager, department heads, and the whole project team.** Obviously, you need to be focused on the development, take care of the team's motivation, solve complex problems, take up challenges, and sometimes work long hours or at weekends.

There are also situations when the project team makes mistakes. It's just the way it is. Sadly, the management or the client all too often and too easily (and unreasonably) accuse people executing the project of lacking in motivation, incompetence, or inadequate management skills. Following that, these people work for years to come with the stigma of "that project," which is continually delayed and costing the company money, while in reality, it's the management team (both on the client's side and on the supplier's side) which were beguiled by the vision of a seemingly easy project and approved jokingly optimistic estimates which turned out to be completely wrong.

During the sales process for one of the larger companies in Europe, the project schedule was worked on, the project in question consisted in delivering a technologically advanced product. The system was objectively complex, it required cooperation with multiple international corporations. The project included, for example, building an additional, intricate module of the system that was missing.

The production manager – based on his extensive experience – estimated the time needed to deliver the product in three years. For the potential client, this was unacceptably long. Therefore, a theoretical model of the new module was designed; this one could be developed in one year. With this assumption the contract was signed. After 18 months, the module still wasn't ready, and at the status meeting, the sales person asked the production manager: "What's the cause of this delay?"

This situation perfectly illustrates the reality in IT organizations. The production team estimates the time of producing the module for three years, after that it experiences a lot of pressure from all sides (sales, management) and is forced to shorten the period to an unrealistic 12 months, and after 18 months (note that the initial estimate was three years) it is asked: "What's the cause of the delay?"

Selecting the Supplier

The life of a project begins during the sales process. In case of extensive transformations led by large companies (e.g., telecommunication companies, banks, airlines), it's 1–2 years prior to signing the contract. The companies, having a lot of experience in supplier selection and management, over the years have developed the right procedures and processes whose aim is to choose the best supplier while observing all adopted ethical standards. The stages of such a process are pre-defined along with criteria for moving across subsequent decision gates. The aim of all of these procedures is to select a company which will be able to offer a product most similar to the expected one at the best (usually the lowest) price.

Larger players have the right purchasing departments (also known as procurement) whose only objective is to manage suppliers. Thus, they're accountable for selecting the supplier, it also includes shaping the selection process in such a way that suppliers offer the lowest possible price and the shortest development time. Purchasing departments' staff are often tough and seasoned negotiators.

The preceding selection process described usually comprises the following stages:

- RFI (Request for Information) – It's a first step consisting in market research. The purchasing department drafts a general request for information and then sends it to a maximum number of suppliers meeting the appropriate criteria. An example could be all companies from the Gartner list which offer the desired products. Each of the potential suppliers prepares a response: the first offer roughly describing the scope, a preliminary schedule, and the so-called indicative price. At the end of this stage, the buyer should have some knowledge about the list of possible suppliers, solutions existing on the market (along with their advantages and disadvantages), as well as a price range of the planned undertaking. Usually the buyer also creates a long list of suppliers that will qualify to the next stage.
- RFP (Request for Proposal) – From the supplier's point of view, it's the key phase of the sales process. It's when the specific shape and outline of the future project are clarified. The client and supplier often hold multiple meetings (workshops) whose aim is to provide the other party with possibly precise set of information on the basis of which a more detailed offer can be created. The existing environment at the client's company is analyzed (AS-IS, legacy); any dependencies between business units, systems, and tools are identified; and the list of responsibilities of the current subcontractors is specified along with their impact on the future project. Finally, users' expectations are quantified. At the end of this stage, suppliers submit very detailed offers. The scope and time should be close to reality. On the basis of the received proposals, the purchasing department narrows down the list of potential suppliers to a short list.

- RFQ (Request for Quotation) – It's the final stage before contract negotiations. The aim of the buyer is usually to optimize the scope and price or possibly find the best combination of both. The potential client already knows the flaws or imperfections of the offered products and is aware of the extra costs that they may expect. He also compares the proposed schedules to the expected one. Suppliers are under pressure to submit aggressive, often risky offers. It's not unusual or even rare that the submitted offers are impossible to execute – just to be in the game and win the client. It is assumed that the situation will develop in one way or another. At the end of this stage, the buyer usually sticks to two or three players with whom he will enter into contract negotiations.
- Contract negotiations – Usually held with one or two players. Most often, the client first tries to reach an agreement with the preferred supplier, and only when this attempt fails (for various reasons), he engages in discussions with the other one. At this stage, the appropriate annexes are drafted: the so-called general terms (General Terms and Conditions – GTC), the scope of the project, project execution schedule, payment schedule, the mechanism of accruing penalty fees are decided, legal clauses, for example, relating to supplier's financial accountability are formulated. The classic part of a negotiation is also an attempt at lowering the offered price to the expected level, which obviously affects the financial condition of the project on the supplier's side.
- Blueprint (also called a preliminary analysis phase or gap analysis – Very often before signing the contract, held at the same time as the negotiations) the blueprint phase is held. It happens when the client is not convinced that the offered scope of the project is correct (thus, the name of the phase – a gap analysis) or his doubts referring to cost estimates (the client wants to demonstrate to the supplier that the project is less complex and therefore the price should be lower). The buyer may ask the preferred player to start a preliminary analysis of the project. It also happens that the client starts the blueprint phase with two players, most frequently one after another.

The process described in the preceding usually lasts between one and two years. Of course, there are some exceptions. It may be easily observed that at the end of the sales process, the following elements had already been discussed and agreed with the client:

- Scope of work – Along with all implementation plans, including the assumed coverage of requirements and the assumed state of the product.
- Schedule – Regardless of the fact whether it's realistic or not.
- Development costs and price – These will determine the future financial condition of the project. If the price is good or at least decent, the supplier will have the means to cover the costs of the development work. Otherwise, there will be problems with acquiring the necessary resources, which means that the project is deemed to generate problems (conflicts with the client, delays, re-negotiations, and arguments about money).
- Payment schedule – It specifies the amounts paid for particular stages of the project, it also decides about the financial condition of the project. If the client agrees to pay adequately high amounts at the beginning of the project (or proportionate to the costs for the supplier), the project is likely to be completed in time. If, however, the client (who believes in his own shrewdness and negotiation mastery) manages to persuade the supplier to accept payment after the completion of the project (post-delivery), he will win only as much that the supplier will be very cautious about the project, will be careful about spending money, which in turn will cause delays and escalation in the project.
- Quality criteria for particular elements of the delivered product – If they're fair, they increase the chances that the project will be executed at a dynamic pace. If, however, they're too stringent, they only give the client (or the users on the client's side) a possibility to reject the deliverables.
- Acceptance criteria for the particular stages.

- Penalty fee mechanism – It's a dangerous tool in the client's hands. If the supplier will be forced to accept too stringent conditions in this area and if the client is determined to optimize costs (and thus uses every possibility to charge penalty fees), the project is doomed to fail.
- Project management methods along with the appropriate plans – Very often they're forced upon the supplier regardless of the fact that they might be completely unsuitable for the supplier's organization.

At the time of signing the contract, all important assumptions are made. Therefore, the sales process is an important, and maybe even a decisive phase of the project!

The Birth of Problems

You could say that the birth of a project during the sales period is like the birth of a child (it might not be the most fortunate comparison, but still). The major changes in the human body happen during the prenatal period. Internal and external organs are created, and the entire organism starts to function. Toward the end of pregnancy, the human body is fully defined. When a baby is born its body does not change dramatically. It mostly grows. It's similar with projects.

The sales process is the hub of all assumptions, but also the source of problems! The subsequent complications are mostly its consequence, and the technical, organizational, or financial problems only deteriorate.

It's during the sales process that the client puts pressure on suppliers to negotiate the best conditions and get as much as possible for the lowest possible price and in the shortest possible time. Suppliers want to get selected so they accept "rotten compromises." They promise that their product meets more requirements than it really does. Suppliers have virtually no choice if they want to win the deal. It's not a matter of one particular supplier who tells lies. In reality, all bidders stretch the truth. Thus, the potential partner is sort of forced to "boost reality" in order to sell the product and acquire a new client. Then they need time to develop the missing features, which results in significant delays.

In fact, the only remedy for this madness is conducting sales processes in an honest and substantive way – that means constructive criticism of bidders' proposals, while respecting their estimates in terms of the effort, schedule, and realistic price. The attempts to outwit the supplier always lead to negative outcomes both for the supplier and for the buyer who will eventually have to pay the price anyway as there will be some extra work to do (changes) or due to the extended time of the project the costs will rise on his side (third party fees, equipment, legacy system fees).

An interesting mechanism is a matter of estimating the price on the seller's side. For example, if bidders are well-organized they involve developers and domain experts at this stage, and, with the help of these people, the costs can be estimated quite accurately. These experts are usually cautious in their estimates. However, the sales process is usually in the hands of people whose main goal is to reach sales targets. In their view, it's better to sell at a lower price but sell anyway and get their bonuses. There is usually a conflict between these groups. The management team in this situation is responsible for navigating the conflict in order to reach optimal but realistic estimates. Unfortunately, the management is often lured by the visions of the sales people and hopes that everything will turn out well, the relation with the client will be good and thus he will pay more, and developers can be mobilized to execute the project in a smarter, cheaper, and faster way. It can't be that difficult after all.

And thus, suppliers are often "forced" to promise that they will deliver the project according to an unrealistic plan (expected by the client). It causes delays at the very start of the project. Such a delay (in relation to a more realistic schedule) is somehow sold to the client. The management team on the supplier's side is aware of this and later has no choice but to exert pressure on the project team to work in the "asap" mode, also often called a "fire extinguishing mode."

A project team usually works at a faster pace than it would be realistically possible but is still unable to catch up with the schedule that was originally sold. It's during the sales process that most delays are created!

It's also at this stage that financial problems have their roots. A common wisdom in the IT industry says: a project that is sold well can be profitable. A project that is badly sold will never be profitable. Sadly, both the client and the supplier trick each other at this stage. The supplier accepts an unrealistic schedule and an estimate that is too low, whereas the client is aware (although it's often not so) that the schedule should be extended and the price higher. Suppliers tend to think that everything will somehow turn out well, that, at the later stages of the project, the client will understand that more effort is required, the costs are actually higher, the schedule will be updated, delays will be taken into account, will pay more on the basis of the Change Requests. Clients are pleased with the deal, they're happy that they bought cheaply and the project will be delivered fast, that they "outwitted" the supplier. And then both parties suffer, projects are delayed, and suppliers have financial problems, which translates into inadequate resources, which only extends the delay.

A project that can end successfully is a project well sold. That means that the scope and time are realistic at the start, and the price is decent.

The most serious delays happen at the selling stage and not at the execution stage! Of course, these are hidden delays that nobody speaks about – yet, everybody knows about them.

To illustrate how the sales process affects the project, let's have a look at the following examples:

In the first one, during the sales process, the supplier offered a very competitive price. It was agreed that some less complex solutions would be implemented. The management team on the client's side approved these plans. During the project, however, management teams on both sides changed. Since then, the client demanded the full scope. The problem deepened because the scope was described in very general terms, which gave the client the power to force the supplier to implement solutions according to the client's interpretation of the system. As, at the initial stages, the supplier didn't have adequate domain expertise he was unable to counter the client's demands effectively. Eventually, the costs on the supplier's side rose significantly in relation to the original estimate. The project was not profitable anymore. The conflict arose between the client and the supplier because the supplier requested – and quite rightly – increasing the budget. The cost of development was objectively higher.

An almost identical situation happened in another project. At the sales stage, some assumptions were made and the scope was simplified. If the client had been willing to cooperate, they would have been implemented. At the same time, the client was a tough negotiator and managed to persuade the supplier to significantly lower the license and implementation costs. Whereas the sold product was not fully prepared for implementation. At the later stage, end users started to voice their concerns and forced subjective interpretations of the requirements, which increased the costs of the project. What's more, the supplier needed more time to develop the missing functionalities of the system. Together with too low income, it was an explosive mixture. The project was delayed, the client was unsatisfied, and the supplier was losing a lot of money.

Both of the preceding cases had some things in common:

- At the sales stage, the client negotiated (and the supplier proposed) the price that was significantly lower than IT transformation costs of similar kind.
- The offered product was only half-ready.
- The scope of the product was very general.

- The client expected customized solutions differing from the product design (the above-mentioned simplification plans had been incorporated).
- The supplier didn't have enough experienced specialists on board who could persuade the client to accept the recommended solutions.
- There were significant delays resulting from this.
- Costs rose exponentially on the supplier's side, thus, in order to reduce losses, the supplier reduced the team size, which resulted in more delays.
- Relations with the client were tense throughout the project.

Summary

- Most people wrongly assume that the project starts when the contract is signed. It's a mistake that can have serious consequences.
- It's said that only some projects end successfully. It's a popular mantra. But we cannot say that most project managers are incompetent or that some typical problems are too hard for the project team to handle.
- It must be clearly stated that only some projects end successful because most of them are sold on the basis of unrealistic plans, unrealistic schedule, and below development costs.
- It's not a matter of incompetent project teams! It's a matter of irresponsible sales process on the suppliers side and wishful thinking on the client's side.

The amount of work to do in the project is objective. Therefore, both sides, the client and the supplier, instead of trying to trick or con each other, should rather focus on the substantive work on the scope of the project, on precise identification of business needs. If the price is too high, instead of bargaining, we should focus on reducing expectations or on devising a plan for simplified implementation. One thing is that the scope should be precisely defined and thoroughly described so that the client can't demand any extra work. Another thing is that the supplier should transparently and honestly estimate the work in order to avoid demanding more money at the first opportunity.

CHAPTER

2

Why Sales People Shouldn't Sell on Their Own

If facts do not conform to the theory, they must be disposed of.

—Maier's Law

Sales is a crucial and, at the same time, a complex area of the business. It's not enough to have the know-how, innovative products, talented engineers, or excellent implementation techniques. In order to operate and grow, a company needs to get through to clients and make them interested in its portfolio – it must sell. There are various approaches and techniques to sales. It must be remembered, though, that sales is just one of the pillars of any business. It's an element that must fit the whole. It often happens that top management is only interested in growth in terms of sales – numbers and charts, while they forget that business growth is not the same as sales growth.

M. Dąbrowski, *Managing IT Projects*,
https://doi.org/10.1007/978-1-4842-9243-3_2

In organizations like these, the organizational culture is gradually disrupted. Instead of dealing with real problems, strategically important activities, working on exploring future possibilities, investing in products, professional implementation, there's too much focus on "here and now" – reflected as pressure to reach sales targets in this particular year regardless of what will happen later. All this happens at the expense of project quality, viability, and product development, building long-term relations with clients is not a priority, which leads to compromised reputation. This in turn results in lower sales volumes in the future.

An IT business which is only (or mostly) focused on reaching sales targets at all costs and does not pay sufficient attention to other areas of the business, a company which does not place enough effort on making good sales offers (accurately defined scope, schedule, and price), in the long run, puts most of its projects in trouble.

There are plenty of examples of companies in which its sales people (in the name of reaching sales targets) are capable of making any silly or ridiculous offer. For instance, clients are offered prices below the costs of work (dumping prices) in order to get the contract signed at all costs. Sales people commit to unrealistic schedules just to get the client and outsmart the competition which may be acting reasonably and lose for this reason. All this is done with the aim to sell, meet own KPIs, and get the yearly bonus. Why is this even possible?

Evidently, the organizational culture of businesses like these is focused on driving sales performance – everybody is turning a blind eye to practices like those mentioned earlier. The decision-making process is set in a way that makes such behavior possible in the name of specific KPIs. Interestingly, these are not small businesses. You can read in the media about some major recovery plans geared toward company restructuring and increasing its profitability (here again too much focus is placed on indicators, only this time different ones). The problem often concerns large IT businesses and even the global giants which have been or were industry leaders.

Unfortunately, if selling like that is possible, it means that somebody is buying and it must be admitted that clients often fall for offers that are too good to be true. The clients themselves can also force suppliers to submit offers like that. As a result, projects last "longer" and cost "more," and funnily enough, these final parameters are actually close to initial and objective internal estimates at provider's side.

The Backstage of the Sales Process – Actors and Their Goals

The sales process is a combined effort of many people and departments. Each element of this constellation has its own goals and specially designed bonus schemes. This in turn determines the behavior and motivation of particular actors taking part in drafting offers.

The sales department – supervises and coordinates offer preparation process (across all departments). Generally speaking, organizations set annual goals for their sales people. These sales targets are a total margin that an individual should get, where margin is understood as a difference between the revenues from the contract and external costs. Thus, in a great majority of organizations, the revenues from contracts are independent of the subsequent expected profit (or loss) on the project. There are both pros and cons to this approach. On the one hand, it may lead to signing of unprofitable contracts; on the other hand, it motivates the production unit of the company to reduce costs and optimize technical solutions. The sales people's aim is to build relationships with clients, learn as much as possible about the prospect's organization, its way of operation, identify decision makers, learn about their preferences and persuade them to buy. They know the buyer's expectations very well. Thanks to that they can significantly influence the content of the final offer.

Pre-sales – these people are responsible for preparing the substantive part of the offer, learning about the client's needs, describing the technical solution, identifying what the product should do, what needs to be developed, and what the scope of work should be. This department is also responsible for preparing the schedule, what may (or may not) be done on the basis of the estimates from the engineering and development departments. They define the scope of maintenance services, including the SLA terms. The pre-sales department usually consists of the most experienced employees, engineering and development veterans. What is obvious, they must also possess good presentation skills. Like sales people, those dealing with pre-sales also have to reach yearly sales targets, similarly defined as company margin (revenue minus external costs) independent of the expected profit. If the appropriate monitoring mechanisms are absent, it might lead to mistakes such as offering unprofitable projects.

Product management – this department focuses on managing products, what includes defining product development plans, roadmaps. Its task is also to ensure that the product is innovative and continuously developed in line with the market expectations, also in relation to the competitors' portfolios. The product management department also makes sure that the offer corresponds to the current roadmap and possibly also prepares a concept of developing the missing elements of a given solution. The members of this department may get their bonuses on the basis of the number of licenses sold, which should help the company to be more product-oriented. They may also be gratified according to the sales targets. However, then there's a risk that product managers will evolve into pre-sales technical consultants, which in turn could disrupt the organization's sales model and cannibalize activity connected with innovations and product development.

Engineering – this department develops products and delivers turn-key solutions, or a mix of both. Engineering managers are extremely cautious, they have experience from previous projects, and they know well what the practical aspects of product and project delivery are like. They seldom believe in simplified solutions offered by the sales and pre-sales departments. They may be driven by the number of licenses sold – then the approach is more product-oriented. More often, however, it's about reaching unit's yearly profit – this approach is worse for the product development but focuses on the business performance of the organization. An ever-present part of the organizational life is a conflict between the engineering and sales departments, for while the sales department underestimates costs, the engineering department quotes the actual or overestimated costs. It doesn't mean that the engineering department is always right. If the product development process is not optimal, competitors may offer lower prices, so it's beneficial when the sales department and pre-sales consultants motivate the engineering department to optimize costs.

Development and maintenance – these departments know both the process of project execution and maintenance of the implemented systems very well. They know a lot about typical problems encountered on projects and ways to solve or avoid them. They can draft a large part of the contract documents, in particular the provisions related to acceptance criteria, the schedule, methodology of work, payment schedule, penalties, and specific metrics in the SLA.

Sales Targets vs. the Future of the Project

As you can see, various departments are involved in the sales process. They have their own annual sales targets and strategic goals (if any) set for them. They also have different bonus schemes, which can lead to a number of problems such as selling unprofitable projects solely to reach the sales targets or offering too high prices (higher than those on the market) to make sure that the annual profit will be made.

Sales and pre-sales departments work under conditions which, to some extent, force them to act aggressively. The organization – which sets ambitious or audacious sales targets for these people in order to grow at all costs – creates an unhealthy work environment in which individuals don't have enough time to develop a well thought-through and safe offer. It's not rare that they need to choose between the strategic well-being of the organization in the future and their own personal good in the moment. If the managements mercilessly evaluate sales people's performance year to year, after some time, those people adapt and learn that they have to sell at all costs regardless of the offer's quality. It's worth pointing out that in general, it's not these individuals' fault – the factors to blame are the bonus scheme and the way it is applied. In the area of project execution, it shows in the following ways:

- Ambitious and unrealistic schedules are offered to meet clients' demands or at least to be better than the competitors, often despite the estimates suggested by the engineering and maintenance departments.
- In order to be able to offer an attractive price, either because it's expected by the client or to be better than other bidders, sales people force the engineering units to approve some abstract simplified solutions – and all this with an aim to get the offer approved within the organization. Consequently, a large portion of work cannot be paid for, not to mention that it will be difficult to acquire internal and external resources for the project. As a result, the project has a low priority in the organization, which means that the less experienced (the cheapest) people are assigned to it. All the experienced people, the organization's intellectual elite, are moved to more lucrative projects.
- To avoid inconvenient discussions with customers about supposedly "unimportant details," too stringent penalty clauses are accepted. Then, clients take advantage of them for months and use them to have a hold on the supplier. In the best case, they use them to get free features; in the worst case, they charge penalties. All this deteriorates the project's situation in terms of finance and resources. Another type of "unimportant details" which are a hindrance to signing a lucrative contract is negotiating the maintenance terms, which means defining specific KPIs and resulting penalties.

The engineering and maintenance units are driven by the department's annual performance. Unit managers get bonuses proportionate to the profit made. Taking this into consideration, individuals focus on making the offers as profitable as possible:

- They try to make the client bear the costs of any missing elements of the product.
- They include in the price the costs resulting from the department's ineffectiveness such as a lacking process, missing CI/CD tools, or lack of the right quality assurance processes.
- They give swollen estimates of the potential work on the project to make sure that they will make a profit even in the worst-case scenario.
- They falsely extend the actual time of the project to have some buffer and also to have the time to complete the team either by hiring people from outside the organization or to wait for the people to finish work on the current projects.

To sum up, on one side we have the sales and pre-sales units which try to make the schedule shorter, lower the price, and agree on unfavorable contract terms to get the deal at all costs. On the other side, we have engineering and maintenance units which fight to make the project profitable, thus the prices they suggest may be too high (in relation to the competitors) and deadlines too long (from the client's point of view).

A Holistic Approach

Objectively speaking, we need an equilibrium between these two conflicting approaches. It's an incredibly tough situation where, on one extreme, we have an easier sales process, and on the other, scope, time, and general terms and conditions that are optimal for the supplier. A solution to this problem is a holistic approach to sales, which means looking at the sales process globally, taking into account the point of view of each area of the organization in order to make the most informed decisions possible.

The first thing that we must have is commitment of each of the "actors" of the sales process. The meaning of the word *commitment* is not limited to being involved in the process, but it also means a personal intention to meet the deadlines at all costs – it's taking ownership. During the process of making an offer, each of the departments involved must be informed and consulted. Product managers must verify the offer against the roadmaps for the products.

The engineering department should estimate the cost and time of developing the missing features or a new system altogether. The aim of the development unit is to prepare a realistic schedule. The maintenance unit should check if the requirements set out in the SLA are possible to meet.

Estimating the work and making schedules is one thing, but the specific departments aren't operating in a vacuum. They have their own business obligations and work in progress, and they need to deal with customers. Therefore, even if the proposed schedule is possible in theory, the actual possibilities to ensure the resources for the project being sold must be taken into account. That's why we need their commitment. To be sure that the project will be delivered, each part of the organization involved must take ownership of their area.

The pre-sales and sales units should not make trade-offs related to the costs, scope, or schedule on their own. At each stage of the sales process, if there is some important strategic decision to be made, an approval of the relevant part of the organization is needed. Only then can we talk about "informed" decision-making. It's clear that the sales people can shorten the schedule by half a year. But it's just introducing a half-year-long delay into the project. It's not going to end well.

Obviously, it doesn't mean that an organization cannot make "bold" decisions, for example, offer an ambitious, even unrealistic schedule or an attractive price – below the actual costs. It can do it to get the deal at all costs, to enter a given region, and to get funding for product development (even if the project will generate losses). It's about making such decisions consciously. It's important to understand how it will affect each part of the organization, which means to be ready for the resulting problems and challenges. Naturally, later the management will encourage the engineering and development teams to work hard. They will do that for objective reasons, so there will be no other choice but to keep the promises made to the client. It's important for the entire organization, though, that the management is acting deliberately and with respect for the employees. That means that project teams are not blamed for those false delays, they are rather asked to make an extra effort in the face of the challenge.

At first glance, it may sound like a soft, sad marketing babble, but in reality, everything depends on the individual's attitude and drive. Literally every employee! People either want to deliver and will do it at all costs or they're feeling disengaged, maltreated, and forced to do ridiculous (in their view) tasks.

A holistic approach to sales should also define the right decision gates in relation to the offer being prepared. Particular managers of contributing units, the manager coordinating the project, pre-sales, sales units, and top management – all of them should have the right tools and data to be able to validate the parameters of the prepared offer. The presented data should be coherent – everyone should be able to see the same numbers and indicators such as the following:

- A detailed distribution of the project costs over time with a division into specific categories (R&D, engineering, development, maintenance)
- A high-level proposal of the project schedule
- Payment schedule, namely, revenue distribution in time in relation to the schedule and costs
- Cash flow on the project (to secure financing of each stage of the project)
- Indicators related to man-day cost and man-day profit
- A list of the most crucial anticipated project risks
- Contract legal assessment (a list of the most important contractual risks, dangerous clauses)

Based on the right tools and data, each part of the organization should have a possibility to either accept or reject the offer. The final decision should always be made by the executives (not by sales people), as the management is aware of the risks then, they also know which departments are able to execute the project and which think that they won't be able to carry out the work according to the current offer.

A holistic sales process should also

- Leave the decisions to a particular hierarchical level, depending on the project's budget. Offers of the highest value should be approved by top management.
- Focus not only on achieving the annual sales targets but also on the profit, it should also be in line with the strategic vision of the company's development.
- Focus on quality, security, and accuracy of the offers. Making profitable offers results in successful projects, good financial performance, employee and client satisfaction, building a good reputation.

It must be remembered that bonus schemes should be designed in such a way that enforces a holistic approach to sales and support a global view instead of leading to selling unprofitable projects with unrealistic schedules. Any attempts at implementing the new approach to sales must be backed by the right renumeration systems.

Summary

- Focusing on reaching sales targets at all costs directly leads to unprofitable and delayed projects (regardless of the efforts made by the project team).
- Sales people should not have a final say with regard to the content of the offers.
- To make an accurate, honest, but ambitious offer, all relevant parts of the organization must be involved in its preparation; in particular, it should be the departments which will be responsible for the project execution.
- The management can make difficult decisions, but they must be well-informed, based on data verified by teams having the right competencies.

CHAPTER

3

Delay As a Problem of the Sales Process

You can't win. You can't break even. You can't even get out of the game.

—Ginsberg's theorem

The reasons for generating delays on the project can be different, but as a simplification, we can divide them into two kinds. First, we have a well-documented classics, meaning delays created during the project execution. Secondly, there are delays that nobody talks about, those that have their origin in the decisions made during the sales process.

M. Dąbrowski, *Managing IT Projects*,
https://doi.org/10.1007/978-1-4842-9243-3_3

Examples of delays of the former type are as follows:

- Problems with the team formation – People must be acquired internally, moved from other parts of the organization; you have to withdraw them from the projects they're currently working on or hire new people from the market.
- Obstacles related to business analysis, not enough right people on the client's side, lacking materials, data quality lower than expected.
- Technical problems – Some tasks turn out to be more difficult and more complex than initially anticipated.
- Individual teams' estimates which simply turned out to be too optimistic, and more time is needed to complete the corresponding tasks.
- Simplified or too general plans which turned out to be based on wrong assumptions from the client's perspective.
- Misalignment between the client and the supplier in terms of the scope, which usually means that the client's opinion is respected and costs rise for the supplier.
- Problems with quality, longer time for bug fixing.

Delays, which are somehow "created" during the sales process have two major reasons:

- Meeting the client's expectations at the cost of subsequent problems in the project
- The need to win over the competitors

In order to win the tender, the supplier often decides to offer unrealistic deadlines based on customer's expectations. Such choices are often dictated by the other bidders. If others accept and adopt such a risky approach, why don't we? What also plays some role here is the organization's overall strategy, goals set for a particular region or client. It was described in the previous chapter.

Delays "Sold" vs. Those Created During the Project

It's really important to differentiate between delays which were included in the contract at the sales stage, before the project even started and those which happen naturally during the project, particularly their impact on the project, on the client's opinion, and the team's well-being.

During the project, there are always some difficulties – more or less serious, but always unavoidable. In most cases, the reasons are to some extent objective and possible to explain, even if someone is to blame (the supplier, client, or a third party). Most of them can usually be kept under control even if they are related to the cooperation between the organizations. Such problems are predictable, they can be explained. There is a lot of time to manage them. Most often both parties are to blame. Besides, the delays caused by this type of problems are not significant, they emerge gradually, in chunks.

On the other hand, delays which are a consequence of the sales process are a completely different story. First of all, they are much larger. We're not talking about two, four, or even six weeks here. More often it's months, a year, or even two. Obviously, this fact is diligently concealed (see chapter *How and when to communicate uncomfortable information or what the client should know*), which is – sadly – a completely reasonable behavior, taking into consideration the future of the project. It's especially true in the beginning when the relation with the client is still fresh and fragile, and the project can be easily stopped at a relatively low cost.

Client's expectation are translated into promises and commitments and these eventually become contract provisions. The discrepancy between the dates specified in the contract and a realistic schedule can be huge at the onset of a project. It's not unusual that the supplier's management informs the client that the estimated delivery date for the first stage is a year later than it was stated in the contract, that delivering on time is virtually impossible. They often suggest "adding new resources" as a solution, meaning increasing the number of people in the areas that are most labor-intensive. In theory, it should help, but not all problems can be solved by adding more developers or analysts; there might also be technical problems which require more time. You must remember that two individuals can't be compared – educating a specialist, especially in the area of proprietary products and business domain, can take years. Projects can't be managed strictly by calculating man-days and FTEs.

From the project team's perspective "the sold delay" can't be rationally explained to the client. The project manager basically has to choose between being a liar and an idiot.

Project managers must either admit that some false promises were made during the sales process (the management will never agree to that) or to act as if nothing happened and come up with excuses, which might be seen by the client as poor management. Thus: you either lied (during the sales process) or you can't manage the project effectively (now)! A negative, even toxic impact of the "sold delays" is unquestionable. They can't be reasonably explained. Members of the project team being in direct contact with the client – analysts, developers, and especially the project manager – usually experience a lot of problems with a communication strategy in such cases. It often happens that they break down and unless they get support from the management – and not only constant pressure and accusations – they simply leave the company.

A good example of this situation is a project for a customer where an unrealistic schedule was sold in order to quickly sign the contract (before the end of the fiscal year) and to reach annual sales targets at all costs. For obvious reasons, the team regularly reported delays from the very beginning of the project. Eventually, after several months, both the project manager and middle management had enough and quit the company.

On one side, the customer was asking logical questions which remained unanswered. On the other side, the supplier's top managers were constantly putting pressure on the team and were blaming them for poor performance. Finally, the customer threatened to cease the contract. It led to difficult but honest talks, "putting the cards on the table," where the actual situation and the down-to-earth planning were presented. When the schedule and expectations became adjusted to reality, the project was delivered with more or less serious obstacles (objective difficulties). The client finally claimed that the project was a success despite some turbulent events.

Delays generated in the sales process can't be controlled in any way. They are inherent to the project. The people in charge of the project aren't able to prevent the root cause of such delays, they can merely try to make up for them. There are no prevention or mitigation strategies, though.

In theory, you can make a risky attempt at explaining the situation to the client. But it's treading on shaky ground. Veterans of project management say: "There's a thin line between honesty and stupidity." Of course, you can tell the client a lot of things, but you need to make a careful choice about what and – even more importantly – when to disclose this information. If we tell the client within one month of the start of the project that it will be delayed by two years, he's almost sure to withdraw from the contract and go to court. However, if despite some problems, we will manage to deliver the first milestone, some part of the system will go to production and only then will we inform the client that the consecutive stages will be delayed, he will – even if disappointed – likely decide to continue the project. There will be some evident outcomes that he will not be willing to lose.

Delays being the consequence of the sales process can't be made up for! It's a futile pursuit of an unattainable goal.

There are two conflicting realities:

- The internal (supplier's) world, a world in which it's clear from the start how big the actual delay is and the expected, realistic delivery dates are (usually) known
- The external (client's) world, where the optics is completely different, where the projects start without any delays and problems will come up only at a later stage

The client's perspective is like this: nothing important has happened since the beginning of the project, so why would the work be delayed? The supplier has to perform some tricks, come up with new reasons all the time, cleverly lay some blame on the client, reach the top of their persuasion abilities.

In a nutshell, reconciling these two perspectives – the client's, who can't understand the reasons for the delay, and the supplier's, who sold the delay during the sales process – is almost impossible. What's more, it's the reason for a large portion of subsequent problems and tensions between the parties. To make matters worse, there are also the natural delays resulting from the problems during the execution of the project.

A classic difficult case in the IT world is a project in which a supposedly already existing product is sold, and in reality, the product is only partly ready or non-existent. The supplier needs a long time to develop the product but doesn't want to admit it (or the management wants to avoid revealing this sort of information). Informing the customer of the actual situation in the beginning of the project could result in withdrawing from the contract and announcing the tender again. That's why the project manager and his team should rationally explain the sold delay from the beginning of the project. It's a challenging and not an easy task, it's complicated from a professional perspective and very stressful as well – both for the project manager and the team members.

Everything boils down to mapping the sold delay to some abstract problems that the supplier supposedly encountered during the project.

First, problems are attributed to difficulties in team formation, then to inability to perform business analysis due to the lack of the right people on the client's side, and finally to poor input data that the client should have provided. After that, there are problems with the design, implementation that all of a sudden proved more difficult than was initially assumed. Finally, the supplier blames the delays on impediments related to effectiveness and quality. At the same time, all client's faults are pointed out and taken advantage of.

Unfortunately, from the client's perspective, it looks as if the supplier lacked project management competencies. How is it possible that each problem is so surprising? Why hadn't the supplier anticipated such situations if such projects are run on a regular basis, one by one? There are some doubts about the project management methodology and standards (the analysis stage). Then the client begins to suspect that the development process has some faults. The supplier's company is audited. The client tries to support the supplier in some way and pays consulting companies for making recommendations. All this leads to raising questions whether selecting this particular supplier was the right decision.

At the same time, the project team is suffering. The team members are professionals and if someone asked them about the source of all the problems and solutions to them, they'd be able to give clear answers about the causes and what should be done next. Sadly, they can't do it.

Thus, delays having their source in the sales process are toxic in any sense of the word! They destroy trust, relations with the customer, people, and the supplier's reputation.

In Practice

Most large IT projects are delayed from the very beginning, meaning the delay is "created" within the sales process. Teams try to meet unrealistic deadlines throughout the project. Delays which had been sold are everlasting in some way. This in turn forces them to work in the 'asap' mode, which leads to people burning out because they just can't cope with the stress connected with constantly having to come up with excuses for delays. It seldom happens that the project team generates such delays at later stages of the project. If they're lucky, the client is understanding and the relations between the teams strong and friendly, then the attempts at explaining the situation might be effective.

Delays resulting from the work on the project can usually be explained and they are not questioned by clients. In short, if we could eliminate delays generated at the sales process, most projects would finish on time. That's why most delays are a problem with a source in the sales process! Such delays are sold with the project.

That's the reason for a huge popularity of agile methods which are based on transparency. They let teams focus on honest and substantive work. When it comes to project management the agile methods is a panacea to a lot of problems. The client must understand the scope of work, must take part in the planning, and set priorities. He should also be aware of the relation between his expectations and the supplier's costs. Progress, risks, and problems are reported and solved on the fly.

However, for the sales people, the agile methods can be problematic. The above-mentioned transparency also causes that the entire scope is clearly visible, and thus the price may be a major put-off for the client. Additionally, clients worry about losing control over the costs of the project, they also think that the supplier will not keep deadlines.

In spite of all their flaws, agile methods are one of the more honest approaches to project management. One thing is that the supplier's costs are covered (the resources are paid for). Another advantage is that the client maintains control over the scope of the project, its priorities, and indirectly also over the costs and schedule.

Some extra incentives may be introduced for meeting the deadlines.

Summary

- During the initial phase of the sales process, suppliers often ignore the realistic schedule to win over the competitors and meet the client's expectations.
- After signing the contract, the project team usually inherits the delay.
- Sold delays are usually impossible to make up for or explain in any logical way. They can't be managed. Thus, these projects are perpetually delayed.

CHAPTER

4

The Scope As a Problem of the Sales Process

If it looks easy, it's tough… if it looks tough, it's near impossible.

—Stockmayer's Theorem

Defining the scope of the project during the sales process is quite a challenge. The process is often complex, complicated, and tedious. It obviously depends on the client, the systems used, work environment, business processes, organizational structure, and first of all, on the particular stakeholders' expectations. The scope of the project also directly depends on the supplier's readiness, meaning the maturity of the products in the portfolio, the level of business acumen, and the number of people available to carry out the work

M. Dąbrowski, *Managing IT Projects*,
https://doi.org/10.1007/978-1-4842-9243-3_4

effectively. It's never accurate to state that "the scope was set out in the contract." All the aspects mentioned earlier can shape it substantially to such an extent that there's a void between reality and the contract.

Defining the scope is a problem which is often avoided at the process of drafting the offer. It's not clearly visible. On paper, the offer seems to meet the requirements. The supplier as an organization signs the seemingly well-defined contract, then kicks off the project which – after some time – always "explodes" for the reasons described previously, which means it expands substantially compared to what was initially planned and that results in financial loss, constant escalations, and delays. This kind of situation is much more common than we think. Finding an effective solution can be hard. The reason for this is that the supplier doesn't want to admit that the scope had been underestimated or that some wrong assumptions were made with regard to the execution plan. The client can interpret this situation in different ways, for example, as a deliberate action, whose aim was to mislead the client. On the other hand, clients may find it harmful to admit that due to clever and subjective interpretation of (high-level) provisions in the contract they are still adding extra features to the scope, meaning that they are getting "more for less." Clients seldom agree for a pre-analysis stage even before signing the contract because they worry that as a result, a much bigger scope would be defined, which isn't in their interest, particularly at the early stages of the negotiation process.

Compliance Matrix

The main tool used by clients to assess bidders (at the RFI and RFP stage) is the so-called compliance matrix. A large portion of the subsequent problems with the expanding scope boils down to the characteristics of this very tool and the way it is approached by suppliers. The compliance matrix is created by the client. It's usually an extensive spreadsheet with a list of detailed questions (or statements) referring to use cases and functionalities of the system to be developed. There might be thousands of questions (or statements) in this document. The problem is that the questions are closed, the supplier can either answer "compliant," "non-compliant," or "partially compliant." There's often space for a brief comment in the form where the supplier can write how exactly the system will meet the particular requirement. Although the requirement list in the compliance matrix is created by users, specialists on the client's side and specifies the actual business needs, the tool's structure and the process of using it can lead to some of the following problems:

- Answers such as "yes/no" or "compliant/non-compliant," "partially compliant" leave a lot of space for interpretation both on the client's side and on the supplier's side.

- The bidder usually has very little time to fill in the form, which makes it a challenging task to find the right people who will analyze the thousands of entries and refer to them within the time limit and in a safe, detailed, and professional manner.
- Describing a coherent concept for a product, platform, or implementation in the form of fragmentary answers in hundreds or thousands of rows in a spreadsheet is almost impossible.

Initially, the compliance matrix is used for shortlisting of suppliers. It's obvious that they are trying to prove their compliance with as many requirements as possible. So asked if they comply, they answer "indeed!" It's really easy to take risks at this stage, each item on the list has a low weight, low development costs. Additionally, it leaves so much space for interpretation that each supplier can always claim that the product he offers meets this particular requirement – you "only" need to know how to use it.

Then the compliance matrix becomes an appendix to the contract. Sadly, later it turns out that the client has their own concept of how he understood the individual items on the list of requirements. And let's remember that there are thousands of them! As a consequence, throughout the project there are conflicts about whose interpretation was actually right. As you can imagine "the client is always right." It's simple: the client approves the system, so to some extent, it's possible to try to convince the client to your strategy of meeting the requirements, but it's usually the client's vision which will win in the end. The results are obvious – a substantial increase in the development costs and delays.

Thus, we end up in a situation where the problem is created before the start of the actual project, during the sales process. The competing suppliers try to prove that they meet all the requirements. They must do everything to win. The scope is defined in general terms; it's based on incomplete knowledge, missing information, and optimistic assumptions; they hope that at a later stage the defined scope will be aligned with their understanding and that they will manage to convince the client.

The Workshops

The better organized and rational clients understand the risk resulting from the imprecise description of the scope. They know that it's in their best interest to inform the future supplier about their business goals, expectations, potential challenges, how their organization works, etc. They want to make sure that the information they give suppliers is complete and allows them to create realistic offers, ones that are based on accurate estimates, and the deadlines in the offers can actually be met. To achieve this goal, clients often arrange a series of meetings, technical workshops, during which specialist teams representing both sides can discuss all relevant issues, including the requirements in each area, environments used by the current systems and subcontractors, integration aspects, migrations, expected or suggested forms of cooperation, management and organizational elements, communication, etc. It's also an opportunity for both technical teams to establish a relationship. It often happens that the client already has some preferences in terms of which supplier he would like to cooperate with – the supplier's organization DNA is evaluated.

During the workshops, the supplier should be able to identify the most important project risks and align the client's expectations with the readiness to meet them. Unfortunately, clients are often not able to devote enough time to each potential supplier. That's why, despite participating in the workshops, there's still a risk of underestimating the scope. That said, it's important to remember that the aim of the workshops is not to replace a project analysis, it's only several, maybe ten-odd days of meetings. Still, they are much more effective than the compliance matrix approach. The supplier can ask open questions here, choose the topics that are most relevant from their point of view.

It's important to point out that during the sales process, the actual project team is usually not formed yet. For this reason, the people taking part in the workshops are most often pre-sales consultants or some (best) engineers from development department. The consequence is that after the project kick-off, there are some obvious problems with sharing these crucial insights with the actual team members. The pre-sales consultants are already involved in new sales processes and too busy for that. The development team gets some partial information. It mostly relates to the assumptions made during the preparation of the offer.

Why the Scope "Explodes"

With time, the supplier gradually discovers what he agreed to deliver in the offer. The scope increases from day to day. The problem is not that the margin on the project shrinks if the company is still making money. However, it's often the case that the supplier makes a loss on the project. The project costs increase two- or even three-fold. It's not the project team's fault. Being able to communicate with the client on a regular basis, analysts don't make mistakes that have such serious consequences as are made during the sales process.

The biggest curse of the projects is a high-level description of the scope.

At the contracting stage, both sides confirm that they understand the subject matter of the contract in the same way. In practice, both the client and the supplier make some internal assumptions with regard to the functional requirements or to the method of implementation. Unfortunately, people signing the contracts are not the same who carry out the work. These new people have their own concepts, not necessarily aligned with the commitments made during the sales process. It often happens that the management on the client's side has some business goals, often conflicting with the users' interest. Some examples here could be any simplifications of business processes, decreased complexity of the client's IT environment, which leads to cutting costs and lay-offs. Thus, the employees on the client's side understand that their position is jeopardized, they're fighting to maintain the status quo, in other words, they are doing everything to copy the current complexity of the environment so that they can work on the new system in the old ways. It quite directly leads to a situation in which the client tries to use the high-level provisions in the contract to their own advantage, which translates into increased costs on the supplier's side.

The problem of the free interpretation of the generally described scope is exacerbated by the tools used for supplier assessment in the sales process. There's a clear difference between the claim that the supplier meets a particular requirement and the detailed implementation of said requirement and the subsequent adoption of the developed solution by end users. Another thing is the simplifying assumptions made by the supplier at the estimation stage. For example, it may be assumed that under some conditions, the development cost of a given module will be 200 man-days (or another unit). If, however, the client consequently boycotts these conditions, the supplier faces a cost equal to, say, 500 man-days. It is clear that if the same problem relates to all areas or modules, the costs will definitely get out of control.

The client's interest is in opposition to the supplier's goals in this case. The client wants to make savings, meaning he will want to get as many features as possible for the agreed price. The high-level scope is a perfect tool from the client's point of view. It may be arbitrarily interpreted by the client to their advantage. The supplier, on the other hand, will want to implement the system in a way that it will be as reusable as possible as a product so that a maximum profit can be made from it and the likelihood of selling it to other clients is high.

Large transformation projects can last for years. In this time, the client's requirements and expectations evolve in line with their business environment. If the project is delayed for various reasons, clients quite rightly expect that the scope will adapt to the current situation. Without any changes in the price, of course.

The supplier is in serious danger when a completely new product has to be developed. It's entering a totally unknown territory. There are usually no domain experts yet at the supplier's company, the competencies have to be developed. The supplier can make a lot of mistakes. The sales process is specifically the stage at which there are a lot of unknowns and thus the scope of the product is very difficult to define. What's more, the client's expectations are not fully understood. Then during the project, the client usually pushes their version of the product concept, which is obviously coherent with the current needs of this particular client. If the supplier lacks the right competencies (which is often the case) and doesn't have the right people, he is not able to effectively protect the business against the client's demands. Analysts are not even partners for the client. As a result, the scope expands according to the client's wishes.

Only by having strong specialists from the very beginning of the project can we use convincing arguments. Without them, we can expect a disaster, a substantial expansion of the scope, and consequently, some extra costs and delays.

The problem of the "exploding scope" also largely depends on the client's organizational culture. The less mature the culture, the more dependencies between the particular stakeholders or departments, the more chaotic the decision-making process, the more influence in the hands of the people trying to achieve their own goals, meddle with the scope, and make the transformation process futile.

In the following, you'll find two extreme examples of the client's organizational culture impact on the project of the same type, the project related to the implementation of the same product in two different companies. In both cases, the client's management wanted to reorganize the users' way of working, increase efficiency, and optimize business processes. It was in obvious contradiction with the system users' interest, who were against any organizational changes.

The first example refers to a telecommunications company located in the Americas. The client's pain point was the lack of hierarchical and clearly defined management structure. In practice, the situation was that on the system's owner's side there was a group of end users that was managed by an operational manager who had no decision-making power. There was no escalation path. On the client's side, there was no person of authority who could manage users' expectations. The supplier declared their readiness for the system acceptance procedure several times! Each time the system went to production and then it turned out that the requirements were wrongly defined by the users and were not able to (or didn't want to) work with the new tool. In the end, the project was halted. There was nothing that the supplier could do.

A similar project in Europe got the support of the top management on both sides. The client set up a formal management structure, with accountable stream leaders, appropriately empowered with the right authority and the real decision making power. The team could report problems directly to the management. They were offered support. There was a special program to prepare the organization for the coming changes. The CTO himself got involved in the project's supervision. With this level of client's company mobilization, the project was bound to succeed and it did.

Apart from the presented mechanisms, there are obviously objective issues resulting directly from bad choices on the supplier's side. There are some whose root cause is the greed or carelessness of some individuals. There are bonus systems that incentivize people to make wrong or morally dubious decisions. These could be, for example, turning a blind eye on the complexity of the project so that the offer seems attractive to the client, deliberate reduction of the costs so that the internal management wouldn't be afraid of the investment and approve the offer. All this with an aim to meet the sales targets and get the annual bonus. It's possible to sign a contract at all costs, show evidence of success and outstanding results, note an increase (in indicators), and then hope that everything will work well the following year.

Good Practices

During the sales process, there are a lot of opportunities to decrease the risk of inappropriately or generally described scope, which in turn can help avoid misunderstandings or conflicts with the client. The first one is the right approach to RFIs, filling any kind of forms or requirement lists such as the compliance matrix. It boils down to devoting the right amount of time for thorough and informed analysis of the questions within. To do it, you need to

- Involve developers, teams that will be responsible for particular areas of the project
- Involve the product management department into the process, these are the people responsible for drafting the concept and product roadmaps
- Take into account the actual implementation readiness during the process
- Adopt realistic assumptions regarding the future development work
- Document all the missing elements to include them in the internal estimation process

Thus, it's necessary to approach the process of preparing the offer in an organized manner. You shouldn't treat lightly the questionnaires that become contract appendices. Yet, since they are just sets of closed questions that you can answer "yes"/"no"/"partly," how can we expect that they don't protect against free interpretation in any way?

The tool that is usually a cure for most issues with an "exploding scope" is a detailed functional description of the product.

Such a document should include the full documentation of all functionalities of the offered product, and more importantly, use cases of particular features which carry out defined business actions. The document describes not only what the product can do (or what it doesn't do) but also how to work with it. The functional description should be an appendix to the contract. Thanks to this, in the event of any conflict regarding the client's subjective interpretation of a particular requirement, the supplier is not defenseless – he can always refer to the functional description and state that what the client bought is precisely defined in the contract, how it works and what wasn't included in the scope. Mature businesses focused on selling products usually attach a document like that to their offer. Interestingly, most businesses don't do that! Preparing the functional description isn't cheap. It's more than just one-time writing of the document. It's a continuous process. Products are a living thing, it's necessary to consciously plan their development. There are investment projects, projects for the clients, and roadmaps. It's necessary to have a team responsible for product management and the right organization of work.

In many companies, it's treated as an unnecessary cost – the organizational overhead! Instead of paying for a team of several people who will consciously manage products, the business tries to save insignificant amounts and then deal with horrendous costs resulting from the poorly defined scope.

When it comes to the workshops organized by the client before signing the contract, it's wise to involve the very people and departments that will work on the project. Firstly, they are the ones who have the experience and knowledge – both about the current state of products' development and the record of previous implementations. Secondly, development and implementation department heads are responsible for managing the resources and know exactly what the capacity is. And finally, it creates a sense of ownership for the subsequent estimates. It's incredibly important!

The team which will carry out a particular task or develop a particular feature approaches the estimation process very seriously because later they will have to meet their own deadlines and complete the work within the budget that they themselves defined!

Whether these realistic estimates will be deliberately ignored by the management is yet another thing. It's worth pointing out that it's always likely that any, even a very experienced person, will make a mistake. They are working on a limited set of data and don't have access to complete information. It's still the sales process and not a final project analysis. The aim is not to eliminate mistakes completely, but to radically reduce their number.

When teams who will carry out the actual work are involved in the sales process, it has another advantage, which is accumulating knowledge about this particular project. After signing the contract, the knowledge transfer is not needed anymore. Even if some knowledge exchange is necessary, it's more efficient, takes less time, and it's also less likely that some important assumptions will be forgotten. An opposite approach is involving consultants in the project kick-offs. After signing the contract, the consultants who signed it join the project team for some time (say, a year). They can work as analysts, product managers, or specialists responsible for implementation. It's very well perceived by the clients, it creates a sense of security that the supplier won't lose the knowledge, intentions, or expectations specified during the sales process. It ensures that the knowledge transfer to the departments responsible for carrying out the work will be controlled and effective. Some clients even demand it as a pre-condition to signing the contract. They understand that their success depends on the right people to a large extent. If the supplier's company introduces this policy, it also creates among the people (in this case consultants) a sense of ownership for preparing the offer. The consultants know that after the sales, they won't be able to "escape" to a new sales process, they'll have to help in the delivery of the solution they'd sold.

A very important and underestimated aspect is acquiring domain experts, especially when the plan is to enter a new, unknown business area.

During the sales process, only the people having extensive experience are able to recognize with some degree of certainty the client's expectations, requirements, missing features, and then estimate the cost of developing them. At some later stages, they will help to define a new product, they'll monitor its development; on the other hand, they will ensure that the scope remains within the agreed limits. They will skillfully prevent and neutralize the client's attempts to extend the scope. They will have a ready, substantive, and convincing answer to all free and subjective interpretations of the contract provisions, they will be able to prove that the supplier's implementation is correct. If you don't have domain experts, you just need to recruit them.

Even if a high-class specialist has to be paid two or three times more than is the average salary in a given position, we need to be aware that if we don't hire him, we will need to be prepared for hundreds, if not thousands man-days of extra project costs!

An extreme example demonstrating the impact of domain competencies is a large B2B project focused on replacing an old IT platform with a new one. The supplier decided to take part in a tender for delivery of the system that was not in their portfolio. The aim was to enter into a new, lucrative business area, acquire knowledge and important testimonials and develop a new product. Unfortunately, the supplier hadn't acquired the right domain competences. Moreover, the team drafting the offer was unaware of its own ignorance. They prepared the offer in which the scope was described very generally. As the product didn't exist, its functional description was non-existent, either, which could protect the company from free interpretation of the scope.

As you can expect, in the compliance matrix, the supplier marked all the positive answers to the questions about meeting the client's requirements. Some ungrounded assumptions were also made. All this was due to the lack of experience in the new business area. Consequently, the supplier submitted an offer which was attractively priced and won, but as it turned out later, it was also underestimated. During the project, the supplier was only acquiring the domain competencies. Gradually the risk of increasing scope and costs materialized. In the end, the initial cost doubled!

It wasn't the project team's fault. It was a problem resulting from the lack of business competencies at the stage of preparing the offer and the initial phases of the project when the specific expectations and way of implementation thereof were discussed.

Summary

- During the project, when the client's expectations are known, we understand the business domain, we have access to the client, the scope may increase, but usually these changes are insignificant, manageable deviations within the profitability range.
- Mistakes made at the stage of preparing the offer have very serious consequences. "Exploding scope" to a large extent is a sales problem and not a consequence of poor project management. It's during the sales process that the most significant underestimations and overly optimistic assumptions are made.
- RFIs/RFPs should be approached seriously and meticulously. A bad contract is a recipe for a catastrophic turn of events in the project.
- Individuals and departments responsible for a project at a later stage should work on the RFIs/RFPs.
- It's necessary to make sure that there is knowledge transfer between the people responsible for the sales stage and the execution stage.
- It's not wise to make savings on the costs of domain experts – neither during the sales process, nor during the project. The losses will definitely exceed the savings. If we don't want to hire two experienced analysts, we can expect that we'll have to acquire tens of extra developers. In order to save money, it's important to invest in experience, knowledge, and in talented people.

CHAPTER

5

Payment Schedule and the Subsequent Condition of the Project

An appropriately sold project is likely to be profitable, but a poorly sold project definitely won't be profitable.

—An industry adage

M. Dąbrowski, *Managing IT Projects*,
https://doi.org/10.1007/978-1-4842-9243-3_5

Projects are not charitable endeavors. Even though it's often forgotten, they're not carried out in the vacuum. Staff generates costs, equipment generates costs, there are the expenses connected with business trips, flights, hotels, third-party contractors, etc. No project can be carried out without thinking about financing, without ensured revenues, without invoicing. You can imagine your own, newly established business which hasn't got any other revenues yet but has just won a tender. If the first payment is scheduled to be paid in a year, after the system has been delivered, how will this new business pay its employees? How will it manage to hire any in the first place?

This problem does not concern only small businesses, start-ups. It concerns all companies, even giants! The fact that they have cash, that they're able to start projects without funding, in some way lending money to the clients, only stems from the fact that these projects are financed with the money earned on the previous undertakings or deals. Even so the problem doesn't disappear. If a large, stable company will not take appropriate care to secure each project, it will end up in similar trouble. Exceptions can be accepted, but they must be some singular cases. Thus, a financing model for the project is a crucial thing.

Planning the payment schedule takes place during the sales process. To some extent, it's at this stage that the later condition of the project is decided, the climate of the project, relations with the client, and access to resources are determined. If the client decides to cover the supplier's costs, the supplier will have the means to form the team, either from the people available internally or on the market. The project will have the right priority. The supplier will be financially secure, so he will not be worried about incurring costs. The work will be executed with the right intensity. The client will most probably be satisfied with the outcome. If, on the other hand, the client decides to pay toward the end of the project, the supplier will be worried about the stability of the contract or the relationship, will be more cautious about spending money, the access to the necessary resources will be more difficult, managers will be careful, they will be making savings – in general, the pace of the project will be much slower than it should be, which will lead to delays and escalations. That's why planning the payment schedule is so important. You need to devote time to it and then try to negotiate with the client to secure the future of the project, try to explain to him how it is connected with the execution of the work and why it's in the client's best interest to plan an optimal and honest payment schedule.

Why Regular Payments Are a Good Thing

A fair payment schedule is one in which revenues are aligned with the costs incurred by the supplier. In this way, the supplier doesn't have to bear the cost of the work done for the client during the project. The payment schedule does not depend on the project management methodology. In a more traditional approach (like waterfall), the first invoices will be linked to the analytical stages, then development stages, various forms of testing, approval, and finally starting the system. In the agile approach, invoices can be issued for time and material, we can also arrange payments at regular intervals, for example, for quarterly releases, which are a delivery of some defined feature packages. It's only important to ensure that the supplier's costs are covered. Of course, it would be perfect if there was also some profit margin on top of that.

Regular payments based on the progress of work are important for the supplier for the following reasons:

- The business generates the costs incrementally and they are covered by the client on a regular basis.
- The business has the means to acquire people and knowledge, to finance the development and implementation.
- Invoices are issued for specific increments, which pushes for continuous progress in the project, introduces some discipline in timely deliveries.

It's also important in the context of people's motivation. Regular milestones and handovers are mini successes which feel like achieving goals, continuous progress. There's no better way to boost motivation than winning. Success breeds another success. Besides, continuous process of invoicing generates all kinds of "team building" budgets, which further boost morale and make the bonds in the team stronger.

Regular invoicing also helps to motivate the management, especially middle management, as their bonus schemes are usually linked to the annual profit generated by their departments. If invoices are issued often, and revenues always cover the costs, such a project is well-received by managers as it contributes to the annual performance and can lead to receiving the annual bonus. It increases the managers' motivation to issue invoices and deliver. It directly affects priorities that they set for their employees (members of the project team).

The project which generates regular payments also has the right PR. If it makes money, it is talked about well. Soon everybody wants to be a part of the project. There's no complaining, there's a lot of interest. Employees want to join the project team. It makes the employees feel appreciated and the project gets competent specialists.

To sum up, regular payments are good for everybody and it's the best possible payment scheme. They're good for the financial situation of the project. Firstly, there are no fluctuations on the project, it's stable. Secondly, regular payments, proportionate to the costs incurred are honest both from the client's and from the supplier's point of view. They're good for the project as there's always funding available for the resources that are needed. They create the culture of continuous delivery. They protect against the pitfalls of bonus schemes such as invoicing upfront with the aim to get a bonus in a given calendar year.

Upfront Invoicing

A common approach during the negotiations is enforcing a situation in which a large portion of the payment for the work is invoiced at the beginning of the project. An obvious and popular technique is selling a license to use the product and invoicing it on the day of signing the contract. Assuming that licenses may constitute approx. 30% of the entire value of the project, it's clear that the invoiced amount would be very high. Such amount can cover several or ten-odd months of the work. Another common technique is distributing the high revenues between the first milestones during the implementation. These can be the deliverables in the form of analytical documentation, first general documentation followed by more detailed documentation, completing the development of a certain module, readiness for integration testing. To a large extent, the aim is to issue invoices for very high amounts, which are much (often several times) higher than the costs generated until the day of issuing the invoice.

Everything boils down to making the client agree to pay more in advance, in the initial stages of the project, and to decide how this bulk payment will be distributed over time aligned with the deliverables.

Clients agree for such arrangements as they often have annual budgets that they must use. Otherwise the management will take away the money that hasn't been used. It's very good for suppliers as

- It secures the financial situation of the business.
- It doesn't only cover the costs of the investment in the project but also ensures high profits in advance.
- The business doesn't have to pay for the work on the client's behalf.

It's also worth pointing out that payments in advance make the client involved in the project in some way. The client who paid a lot in the beginning, has made an investment. Therefore, it's not easy for him to resign or withdraw from the project. He will try to do anything possible for the project to succeed, otherwise the money he invested would be lost. This in turn leads to a proactive approach on the supplier's side. It is clear that the client is serious about the project (by paying a lot in advance), and the supplier starts work at a fast pace.

In theory, issuing invoices in advance is good for the project. If a business makes money from the start it can not only cover the costs but also demonstrate high profits, which obviously leads to giving the project a high priority. The rule is simple: higher revenues, more competent people. So the project has access to the right people and means from the start. The project is also well seen in the company. Succeeding in a project like that is a chance for a promotion, proving one's value for the company, etc.

Unfortunately, there are also some negative aspects to this type of payment schedule. One of them is the urge to show profits quickly. Instead of securing the financing for the subsequent stages of the project, the management decides to use (recognize) the revenue as quickly as possible, in the current calendar year, also in advance. It usually depends on the financial needs in a given year. The business can be losing money on other projects and the invoices issued in the new project may be some way of compensating for these losses. It can also be a way to show a higher value of the business. There are a lot of motives to act in this way. Problems can also be generated at the middle management level. Depending on the bonus schemes, managers may be tempted to show high performance in order to earn high bonuses at the beginning of the project. All this in the vein: "You never know what the future brings, it's better to get your bonus here and now." It leads to a situation in which funds for the project are "used up." It can be a deliberate approach, but it can also be done without the knowledge of the top management.

- Deliberate action is undertaken when a business wants to show better financial performance.
- Secretly – when the middle management shows high financial performance (due to invoicing in advance), but at the same time, hides the actual status of the project, cleverly hiding the costs that will have to be borne in the future.

All this negatively affects the work. Due to the fact that most revenues had been invoiced in the beginning of the project from an accounting perspective, it will be generating losses in the following years, despite the total profit which was shown in the very beginning! This in turn will lead to cost cutting in order to minimize losses. As a result, there will be delays, escalations, and a general deterioration of the project. All of this is a consequence of companies focusing on yearly financial KPIs and profits. In the subsequent years, the management tends to forget that the project is overall profitable, that high amounts had been invoiced in the first year.

To sum up, planning a payment schedule in which most revenues are invoiced up front, in bulk in the beginning of the project makes perfect sense. It can be beneficial for the business, the project, and the client. It can significantly enhance the level of the client's and the supplier's commitment, increase the pace of the project, and boost the team's morale, etc. It must be kept in mind, however, that it entails risks that we must be aware of and these must be effectively managed.

Aggressive Payment Schedules

There are situations in which clients use their bargaining power to the limits. In the very beginning of the sales process, they inform that the prerequisite to go through the final stage is that the supplier agrees to the method of financing they find convenient. The very fact of adjusting isn't the main problem here. The thing is that these clients enforce a model in which the larger part of the payment is made only after the final delivery of the system to the production or even two or three months after that, after going through the so-called stabilization period, achieving some business goals, etc. They often suggest a success fee, a bonus for getting certain benefits. Such clients speak in the vein: "If you believe in your delivering power, you're sure you'll deliver, you have nothing to worry about." You could obviously question that, you could find counterarguments. There are also clients who want to have an option to withdraw from a project at any time, they're not sure about the results, so they don't want to incur any unnecessary costs.

Aggressive payment schedules are harmful for the projects! In theory, they provide clients with control mechanisms, they give them an option to withdraw from the investment, but if you think about the outcomes that you want to achieve at the start of a project, the result will be quite the opposite.

Here is why. Firstly, in such a project, the supplier bears the costs on behalf of the client, so from the supplier's point of view, the balance is negative until the system goes into production and gets the final approval. The supplier can't withdraw during the project as this could generate a loss – only the supplier is bearing the costs at this point. At the same time, the client has all the tools to enforce extra work, increase the scope, blackmail. If the supplier doesn't meet their expectations (even those going beyond the scope of the contract), the client can easily terminate the contract, withdraw from it. There had been no significant costs on their side. Even if the supplier goes to court, he will have to fight to get the invested money back, not to mention about lost benefits for years of work. The contract is bad for the supplier in this case. For this reason, in the projects with an aggressive payment schedule, suppliers usually act with great caution. It's clear for them that the client isn't bearing any costs connected with the project, so they're afraid to spend a lot of money in the beginning and then they also try to strictly control the costs, their activity is not performed at full capacity. As a consequence, the project doesn't have access to sufficient resources, there are not enough good people involved. Within the supplier's organization, department managers don't want to take risks and they rather assign competent people to other projects that are safe from a financial perspective. Finally, there are delays and escalations, each of them being a threat to the project. Each serious problem can lead to the termination of the contract. The client has lost nothing but their time.

That's why during the negotiations, the supplier should avoid aggressive payment schedules at all costs. It's in the best interest of the supplier, the project, and the client himself.

On-Premise vs. SaaS

The terms on-premise and SaaS are common in the context of offers. On-premise is a traditional offer in which the client buys the product, including the implementation and customized installation of the system. SaaS (Software as a Service), on the other hand, in theory means that clients buy a service which is an access to the product – it's important that the system is available and efficient, the supplier's responsibility is to maintain the right hardware infrastructure and the access.

When it comes to large transformation projects, the implementations of B2B systems, the terms on-premise and SaaS are mostly used with reference to the financing model from the client's point of view. The reason is that they expect a customized version of the system anyway. Large companies act in a characteristic way, they have their specific requirements. These are difficult to meet with a product approach like SaaS. Such clients have specific business processes that they can standardize only up to a certain point. They have their unique products that they want to maintain in a way that is different from the competitors. For such clients, the on-premise option simply means that the service will be financed in alignment with the supplier's costs. The SaaS option means that a lot of costs will be incurred only after the implementation and start of the system, during the maintenance period. So, it's only a matter of optimizing costs depending on the requirements that the client has with reference to cost qualification, that is if the budget type is CAPEX (Capital Expenditure) or OPEX (Operational Expenditure).

For the client, the on-premise model is beneficial when there is a large CAPEX budget for investments. The SaaS model, on the other hand, is selected when there are OPEX budgets, meaning that a lot of money can be spent on system maintenance. As you can see, everything depends on the client's accounting policies.

For the supplier, the SaaS offer is a good option only when the model in which he operates is really SaaS, when the portfolio consists of existing, mature products with a well-defined functionality that is useful for an entire target customer group. Implementing such a system is cheap, fast, and hassle-free because it's the new client who has to adapt to the new product's functionalities. Clients can express their expectations and the supplier might put them on the roadmap, but only if they're in line with the product development strategy.

When a large transformation project is sold and we agree on a SaaS model, when it comes to the payment schedule, we need to keep in mind that it's just an aggressive payment schedule and the client is just trying to avoid the costs of implementing the system, he wants to postpone them to the maintenance period. The reason for such client's behavior is clear: these are some deliberate financial optimizations and not leaving a way out of the project. Thinking about the future of the project, it's important to try to make the payment schedule aligned with the cost distribution in time. Only then will we be able to create a sufficiently stable work environment (as was said before).

Summary

- Regular payments are good for the business, the project, and the client.
- An optimal payment schedule is the one in which invoicing is aligned with the progress. The supplier's costs are covered on a regular basis. A culture of continuous delivery is created.
- Invoicing in advance has a lot of advantages, but it also entails risks which need to be understood and deliberately managed.
- Aggressive payment schedules are harmful. They should be avoided at all costs. They create a toxic work environment and poor relations between the companies. They make the supplier act with caution because the client can terminate the contract at any moment without any financial consequences.
- During the negotiations, most emphasis is usually put on the amounts, especially total revenues and profit, but it's equally important to look at how these payments are distributed in time.

The payment schedule is not a table that sales people make ad hoc! It's important to stress again that the sales process largely determines the condition of the project, the work environment, relations with the client, access to resources, the reception of the project within the company, and in particular it determines the supplier's attitude throughout the whole project.

CHAPTER

6

Why Attitude Matters More Than Outcomes

Attitude is a small thing that makes a big difference.

—Winston Churchill

It's a common conviction that in project management, an outcome is all that counts. Everybody is focused on achieving goals. The scope must be delivered according to the contract, approval protocols must be obtained, which means that the proper quality was ensured. It's also necessary to be within the budget. As a matter of fact, most businesses focus on time and budget, meaning that they want to deliver in the shortest possible time and as cheaply as possible! Outcomes must be delivered at all costs, also at the expense of the client, what often results in a constant conflict with the client lasting throughout the whole project. The disputes are about who is guilty of this or other delay, pointing out each even most minor fault in order to lay all the blame for the problems on the other party. It's necessary to show the outcomes or at least not to be responsible for the lack thereof! Then there are also the provisions of the contract, reading the contract very literally, in particular to stick

M. Dąbrowski, *Managing IT Projects*,
https://doi.org/10.1007/978-1-4842-9243-3_6

to the costs laid out in the contract, but also to point out the mistakes of the other party. The question is: what's the point? Acting in this way, we focus on finalizing the project at all costs, also at the cost of losing the client. It's possible that the project will be delivered, not only within budget, but also making the client pay extra. What will be the outcome of this behavior, though? It will be our last project with this client because our arrogant, uncompromising attitude will stick with him and he won't be willing to have anything to do with us in the future!

We should rather care about building long-term relations with the client, meaning that this particular project will only be a prelude, a first step. To do that, we should show the client that we are his partners, we understand him, we want to solve his problems, even if it means jeopardizing the outcomes of this particular project.

Performance Is Not Enough

Outcomes are extremely important, but not the most important. If the project generates tangible results, in theory everything should be alright. The plan is executed properly, necessary changes are incorporated, implementation of the new system is on time, so the client should be pleased. He's getting exactly what he'd paid for. The outcomes are also the supplier's advertisement as such. They are proof that the work is being done well, deliverables are provided, the supplier is effective.

However, it turns out that in reality, it is not as obvious as it may seem. It happens that the client stops the project despite "good" outcomes. Somewhere at the top, decisions are made about mergers, acquisitions, or selling the company. New management comes along with their "befriended" suppliers.

It can also happen that despite fulfilling the provisions of the contract the supplier is not able to reach an agreement with the client who is disappointed with the cooperation style. The supplier keeps pointing out client's mistakes and brings attention to any fault in order to cover his own delays. Besides, the supplier is not flexible, even the minutest deviation from the scope in the contract is treated as a starting point for negotiating some extra remuneration.

Usually the client's business changes and evolves with time, the project's conditions change, it is clear that the modifications planned by the client were not well designed, some things were not taken into account or thought through, and finally, the customer is trapped by the unsuitable contract. Objectively looking, there are significant changes in the scope and approach to the project that the customer desperately needs to introduce. Nonetheless, the supplier is mainly concentrated on charging the client and delivering the contract to the letter – word for word.

Very often the client has his own problems with the project execution. There is not enough people having the knowledge about the legacy systems that will be replaced. He's not able to process the source data. There's not enough knowledge to prepare the input for analysis. Another problem might be the lack of knowledge about the database patterns of the current systems as the people who created them are long gone. There might be trouble with mapping the interfaces between the current systems. The client's problems are therefore objective. The supplier could help, but doesn't because, in this case, all the delays in the project are the client's fault and the supplier shows that he delivers on time.

From the client's point of view, delivering outcomes according to the signed contract is not a benchmark of project success! Neither does it generate the client's satisfaction!

An example perfectly illustrating what's just been said here is a project conducted for one of the biggest companies in Asia. The supplier signed a contract for developing and implementing a software product. The contract clearly defined the functional requirements of the product, and – what is more important – contained a detailed description of the scope with regard to the implementation and configuration on the supplier's side. After signing the contract, the product was delivered and configured exactly how it was specified in the contract. According to the supplier, all provisions of the contract were fulfilled, and deliverables provided. Yet, during the course of the project, it became clear that during the sales process, the client misunderstood the configuration capacity of the system and consequently incorrectly estimated the scope of work to be done by the supplier.

As a result, the system was useless for the client. For obvious reasons, he didn't want to pay for the work, as from his point of view, the system did not work. There was an open conflict between the parties fueled by an exchange of formal letters. It nearly finished in court. Fortunately, on the supplier's side, a new management team (including a new project manager) was introduced and they adopted a more client-friendly approach. They took the time to talk to all stakeholders on the client's side, during which they tried to understand the scale of the problem and the actual needs regardless of the provisions of the contract. After another year, the system was formally approved and enabled. The supplier had some extra costs but gained another testimonial for the product, a new happy client who eventually paid everything he owed.

As it turns out, blind execution of the contract was not enough. The decisive and crucial factor was the new team's attitude, an attempt to understand the client, the willingness to help in meeting customer's business goals.

The Power of Attitude and Devotion

What ensures the greatest safety in the project is not the contract and its provisions. It's not delivering on time. It's the relation with the client and the image that the supplier creates in his eyes!

Such image is earned gradually, slowly, mainly in crisis situations when both parties encounter problems. In such moments, the worst thing the supplier can do is to try to lay the blame on the client, demanding money when the client is cornered and has no choice. The best solution in such cases is demonstrating a customer-oriented attitude, showing devotion, the willingness to solve problems. It's definitely not about working for free. The attitude is what counts. First, we show the willingness to help, present some possible solutions, and then signal the readiness to act. Sooner or later, the client will pay these extra costs, it will usually be his own initiative.

The first element of shaping a customer-oriented attitude is making an effort to understand the client!

Contrary to what you might think, it's not that obvious. Most often, the project team focuses on how to fulfill the provisions of the contract, which is not the same as pursuing business goals that are important for the client. In practice, the client also makes mistakes in the sales process. He also makes some assumptions, often wrong, based on incomplete data or too superficial analysis. That's why, from the very start of the project, the supplier should focus on understanding the real needs of the client, regardless of the scope defined in the contract. As a result, in general, the scope of work is finally extended, which benefits the supplier in the form of extra profits, but only if the contract is properly formulated (see: previous chapters) and the supplier's attitude is proactive.

First of all, for the client, a supplier who understands him and can see things from his perspective is an ideal partner. Apparently, the clients are more annoyed by a lack of understanding than by delays!

Clients are aware of the fact that problems are a normal aspect of every single project and that projects gradually become more complex. Lack of understanding, however, is a serious problem for them! If the client feels that his problems, suggestions, and enquiries are ignored (for some reason), he soon comes to a conclusion that regardless of the outcomes so far "this supplier is impossible to work with." Then the client loses motivation and the willingness to continue and as the project is difficult, requires effort and determination, it may simply be stopped.

Secondly, a very important element is the genuine willingness to solve problems, especially those objective ones related to both parties and client's problems!

Such an attitude must be presented at all times – during each conversation, during formal and informal meetings, in email communication, or formal letters. After some time, the culture of helpfulness and problem-solving becomes natural for the team and the client's trust increases. Thus, a work environment is created in which people, instead of fighting, focus on the work. If, for example, the client's knowledge about legacy systems is not sufficient, we should suggest that we'll take this responsibility instead of using it to cover our own delays. If the client has staffing problems or lacks certain competencies in the area of data migration, for example, doesn't know enough about database patterns of the current systems or does not have the people who could deal with them, we can either wait and accuse the client of delaying the project or help and take this responsibility.

Of course, it does not come for free. It's always some extra revenue, but sadly, it comes at a cost of losing one's position. By taking ownership of all the aspects of the project, the supplier also takes the risk that he will become the only source of delays. That said, if the client can see that the supplier acts in a pragmatic way, helps, takes ownership, he very seldom blames the supplier even in case of problems. The client understands perfectly well that he would also have problems with the undertaken activities. If the supplier solves his problems, cares about them, then even in case of delays, the client will not start looking for a new supplier. First of all, he will become so dependent on the current one that it will be virtually impossible. Secondly, where will he find a supplier with a similar attitude and knowledge about his systems and work environment?

Suppliers often worry about taking ownership of areas which are problematic for the client only because they're afraid to lose the comfortable and secure position. It's a serious mistake. If there are objective problems in the project, even if they're caused by the client, the client's management might be forced to stop the work.

We shouldn't feel schadenfreude about the delays on the client's side. We should be afraid of them! It is dangerous for the project and negatively affects payment schedules. That's why the best solution is to offer help and take ownership for customer's problems.

It's obviously a paid service – the client would incur these costs anyway. The supplier doesn't lose his position, quite the opposite, he gains a reputation as he makes the client dependent on him. If the client is happy, he will not punish the supplier, even in case of technical problems or additionally generated delays. If the client feels that this is the only supplier who will always try to help at all costs, the client will be willing to put some more money into the project, especially when it turns out that the supplier's costs are higher.

Solid outcomes are not the most important thing. Projects have their ups and downs. There are some tougher periods, downtimes, delays. There're also successes. Nonetheless, the supplier should always try to understand the client and support him in solving problems. Such an attitude is the only constant, it's a foundation, on which the future (not only of this particular project) is built. It determines the whole (current and future) relationship with the client.

The example described earlier was about a difficult implementation process for a client in Asia. For contrast, we're going to look at an implementation of the similar product in a different company in Europe. In this case, a dedicated team was appointed, its members worked onsite at the client's premises. The team knew exactly what the client's needs were. Individuals on both sides had friendly relations. The project was conducted in a proactive, flexible, and transparent way to such an extent that the client had complete knowledge about the development plans of the supplier's product, he understood and respected them, and even actively supported them, while trying to avoid customer-specific solutions that would benefit only this particular client and jeopardize the supplier's product's roadmap at the same time!

The Project vs. Relations with the Client

All this shows the importance of the right attitude, a proactive approach. The client does not remember that this particular milestone was delivered. He will tend to remember that when it was tough, while facing huge adversities, we were always on his side, he could rely on us, could count on us to suggest a solution regardless of the contract. The client will remember that it's worth talking to a supplier like that.

Additionally, individuals on the client's side will also remember this. They will become our best advocates! After some time, they change employers and it turns out that suddenly we get an inquiry from a new prospect and all this happens because someone remembers they worked with us before and has good memories from this cooperation.

The project itself is just one aspect of shaping the relations with the client. It's often only the first step. Then there come services such as maintenance, the system is developed, new features are added. The client may decide to buy other products from the supplier's portfolio and thus have new projects.

The client may forget this particular project, but he will always remember how he was treated by the supplier. If he was constantly under pressure from the supplier, if there were some tricks and manipulations, if the supplier stuck to the contract uncompromisingly, escalated, or demanded. Or maybe he was friendly, demonstrated a positive attitude, always tried to help, suggested solutions without blaming anybody, adapted to the situation – all this so that the client can achieve his goals.

Even if a particular project failed for some reason, even if it was difficult, the client will remember that the cooperation with this particular supplier was good and will choose him again at the first opportunity. The supplier will be asked for an offer for a different product from his portfolio and will stand a good chance to win.

The aim of the example presented here is to show that a supplier can lose a client and prospects for a lucrative contract due to myopic thinking and maximizing profits. The case relates to a major transformation, replacement of some key IT systems in South America. The project started at a dynamic pace as the client had a high budget and decided to pay a large portion of the costs up front. In such circumstances, the supplier instantly gathered a strong team, was proactive, tried to solve the client's problems, manage expectations at the local branches and synchronize the requirements with the HQ. For the first two years, the project was run in a customer-oriented manner, teams cooperated in a friendly atmosphere. The supplier's performance was high. The client was very happy.

A serious impasse occurred after several years of cooperation. On the one hand, the client was planning to start the system in production and encountered multiple organizational problems mainly stemming from the scale of the business, poor management structure on his side, lack of willingness for migrations, local problems with convincing users to start using the new systems, but also from the delays on the supplier's side. On the other hand, the client was running out of money, but at the same time, he expected even more commitment and more flexibility. The supplier righteously asked for increasing pay as the scope had become much bigger than what was written in the contract. The client, not being able to see the business value, refused to pay more. The supplier then forced the client to increase the budget.

From the client's point of view such a negotiating position was arrogant. The supplier won the battle for more profit but lost a war for the relation! At the same time, the supplier started acting more stiffly. He tried to match the costs to the revenues, which ended up in lay-offs, moving the best people to other projects. The initial customer-oriented attitude was replaced by a stiff approach: constant bargaining, executing the work according to the provisions of the contract.

For the next few years, the project was conducted in terrible conditions. People worked under pressure, they were demotivated, and the client wasn't able to manage expectations and coordinate the work on his side. The relations between the businesses deteriorated from day to day. Eventually, the systems were mostly implemented, but nobody wanted to use them. The client decided that he would never have anything to do with this supplier regardless of the fact that most of the contractual obligations were fulfilled.

We could ask the question: is it more profitable to fight for this particular project and pay the price of losing this relationship or perhaps it would have been wiser to invest in maintaining good relations with the client and stand a chance of doing business in the future?

Summary

- Executing the work according to the provisions of the contract, getting solid outcomes does not guarantee the success of the project. They might not be sufficient from the client's perspective.
- For the project to succeed, it's necessary to have a good attitude, be customer-oriented, understand the client, and show willingness to solve the client's problems.
- If the client feels understood, respected, and treated as a partner, then despite problems and delays, the client will be pleased anyway and will not punish the supplier.
- The project is just one stage in working with the client. We must always remember that the relation with the client should be the most important thing. It is a guarantee of the future possibilities, new deals, and extra profits.

CHAPTER

7

Why Being at the Client's Premises Is More Important Than the Outcomes

Get closer than ever to your customers. So close that you tell them what they need well before they realize it themselves.

—Steve Jobs

M. Dąbrowski, *Managing IT Projects*,
https://doi.org/10.1007/978-1-4842-9243-3_7

In the business realm, we often witness situations in which average companies succeed, where, at first glance, these companies are often worse than their competitors. On the other side of the spectrum, there are businesses which have excellent or just better products, hire better specialists, their business acumen is impressive, they are proficient at carrying out well-defined tasks, but somehow they don't succeed. Their projects are stopped and they are replaced by companies which are weaker in just about every aspect. These supposedly worse companies don't have proprietary products, or their portfolio is very limited, the solutions they sell require a lot of engineering effort and production resources during the implementation. However, they take over the business, carry out projects effectively, and become partners for years to come.

Why is that? The answer is that, in general, all companies hire people, not robots. And people are seldom objective. Their perception is affected by multiple factors. When it comes to managing IT projects, clients assess progress in a very subjective way. They have their opinions about the quality of the cooperation, talks, meetings. They build their own opinion about the supplier's competence, his maturity, and reliability. They also assess the systems being developed subjectively. To make matters worse, they also have their own subcontractors who want to take over our business. Thus, they also try to show our company in a negative light.

So, can we influence how the client sees our company? How can we make sure that the outcome of our work is recognized and appreciated?

What can we do to make the client fully aware of the efforts we are putting in the project, to make him appreciate our work, our willingness to solve problems, or the advantages of the system that we're developing?

The answer that comes to mind first is communication. In theory, communication is crucial, its lack is the main problem at all management levels. In this chapter, we will look at a very specific area, and that is how the local presence at the client's premises can contribute to the project's success. Let's look at the advantages of this approach, what problems it solves and why it's always a worthwhile investment on the supplier's side. We will see in particular that a team present locally can make up for a lot of deficits and the client will always tend to choose a supplier with an on-site team rather than the one having a great product but working remotely or seldom visiting the client on-site.

To answer the previously asked question: companies which don't have excellent products or world-class business acumen can significantly increase their chances of success in their projects when they invest in being present at their clients' sites. It's always seen as a form of dedication, a step toward a long-term partnership. The customer thinks in this vein: "It's not a supplier who will only want to make money on the project and then he will disappear leaving all the problems behind. This supplier thinks long-term. He will stay for longer. He is interested in building a relationship and new areas of cooperation."

The Added Value of the On-Site Presence

Forming on-site teams at the client's premises has a positive influence on multiple aspects of running a project. It also creates the right image of the supplier as a proactive company and a reliable partner. First of all, it helps to build relationships with particular people on the customer's side, often with those who are not directly linked to the project, but whose opinion counts. In this way, the supplier can

- Identify all communication channels, key people having influence on the project
- Build credibility and gradually win some trust
- Shape one's image in the eyes of the client
- Actively manage how the progress in the project is perceived by the customer
- Effectively influence the decision-making process at the client's side
- Defend the business against hostile activities of the other suppliers present at the client's premises, and of some individuals within the client's organization

In theory, the client should appoint some people accountable for specific areas of the project in the very beginning. However, during the project, it often turns out that the management structure on the client's side isn't optimal, does not represent all the stakeholders, it has some weak points. There are situations in which the people working on the project have no decision-making power, must consult with other departments, in fact they need to discuss decisions with some influential people who are not a part of the project. Such people's opinions really count. They can even decide if the analysis is successful in the first run or make their negative statement one year after completion of the analysis, when the system is already in the production stage, but it turns out to be incompatible with users' needs. Such influential people can be the decisive factor for the supplier's reputation.

That's why it's essential to identify who the "major players" in the customer's organization are, who will be actually using the system, what is important for them, how to meet their expectations, etc. It's also important to find out about the information flow at the client's organization. Who communicates with whom, what for, how often, how significant decisions are made. All of this should be determined regardless of the project management structure that the client proposes or of who the appointed contact persons will be. It's clear that such an analysis is impossible to make remotely, it's very difficult to do it on an ad hoc basis by just sending someone over from time to time. The situation is completely different when we have a team working on-site, the team that somehow blends into the client's organization. Then relationships are built organically. Gradually, day after day, the supplier's representatives have a chance to uncover the inner workings of the client's organization. In this way, a lot of mistakes can be avoided as a better understanding of the client's business is achieved, and every information can be verified with a larger group of people.

Thanks to working with the customer on a daily basis, having built relationships with particular individuals, the supplier is even capable of performing certain tasks for the client. The supplier asks fewer questions, can figure out a lot of things himself, can proactively suggest solutions to problems that the client himself might have overlooked. In this way, more trust toward the supplier as a partner is created in a systematic and natural way. His credibility grows. The quality of the cooperation is enhanced. It's easier to solve problems, information is gathered more quickly. This in turn leads to shorter time needed to do certain tasks.

We must also remember that the progress of the project is seen from a subjective point of view. It's easy to say that something hasn't been delivered. It's usually obvious: a milestone has been delivered or not. However, what's crucial is the context. It's one thing when the supplier hasn't appointed enough people, hasn't set the right priorities and for various reasons hasn't been working on removing some impediments and waited idly to see how the situation develops, but it's a completely different thing when lack of progress is a result of some objective problems, including those which were on the client's side, while the supplier was doing everything to solve them. The thing is that the client, especially the management, might not be aware of the situation.

It's not easy to break news like that. The on-site team can have significant influence on how the supplier is seen and how progress is presented. The local team can inform their partners about problems, challenges, progress on a daily basis. The information spreads in the client's organization through some informal channels of communication. On-site you also have access to the management. You can ask for an informal meeting or even arrange to hold such meetings regularly to have a chance to discuss challenging situations, actions taken, and possible options, you learn their opinions, get advice, and so on.

What's more, being at the customer's premises you can actively influence the decision-making process. Such decisions are often made during the informal meetings. The client is always open to listen. For a manager, it's not a problem at all to spend 15 minutes during a coffee break to listen to the supplier, especially if a serious problem can be solved in this time. Thanks to this time investment, making decisions at the steering committees is a pure formality. Of course, in order to have that coffee you need to be present on-site.

It's similar when it comes to defending oneself against the hostile activities of the competitors, other suppliers, or some hostile individuals from the client's organization. In order to prevent them, you have to be a part of the client's organization, you need to be aware of the rumors, of what is talked about. It may sound odd, but it's extremely important. You can often anticipate some problems ahead as they don't emerge overnight. Relations with particular individuals, informal meetings, being present on-site are of key importance here. In order to effectively neutralize this sort of threat, first you must be able to see them. It's common for suppliers to blame one another. Each of them tries to show themselves in a positive light, presents the situation from a subjective perspective. The one who has access to the top management wins. A word spoken during a coffee or a lunch break is enough to change the client's attitude toward others. That's why it's crucial to be present on-site.

On-Site Presence As a Business Model

At escalation meetings, lack of the supplier's on-site team is often quoted as a primary reason for all sorts of problems. It relates in particular to international projects carried out in different countries. The situation gets even more complicated when the supplier and the client have very different cultural backgrounds. We encounter a lot of different barriers, including the language barrier. Our value systems vary and so do the way we act in different situations. Nonverbal communication and body language are not the same. Hierarchy in the workplace is seen in other ways. There are some unspoken conversation rules that we don't understand. The approach to solving problems is different. There are a lot of differences and numerous books have been written about them. In this chapter, we will only focus on the aspect of on-site teams.

From the client's point of view, the unwillingness to invest in the local presence may lead to trouble. Customers have often problems organizing themselves and expect that the people on the supplier's side will understand their company structure, will figure out who to talk to. They expect that the supplier will ensure preparing the correct and complete stakeholders' map, that all of their needs will be correctly identified. When there are problems with the progress of the analytical stage and scope definition, clients will usually blame it on the lack of on-site presence. Even if it is their fault, they will be putting pressure on the supplier to form a local team. Often they act like this on purpose. From an objective point of view, on-site presence solves a lot of problems, primarily those referring to training the users on the system's functionalities, instructing them how to operate the system in the post-production phase when users need constant assistance and don't know how to use the new tool.

In international projects, if the customer has to choose between the domain knowledge and the on-site presence, in most cases, he leans toward a supplier who offers on-site presence and a local team.

If there are two subcontractors working on one project and there are some complex problems to solve between both of them, it's always the one who works on-site who has better relations with the customer and finally puts all the blame on the other subcontractor. Then even if you're in the right and deliver everything on time, it doesn't matter unfortunately. What counts is the client's perspective. And this perspective is controlled by the company whose team works on-site! They are the ones who have access to all the decision makers.

Forming on-site teams and competencies adds so much value that some businesses base their business model on that. They specialize in effective search and training of large number of professionals. They can quickly build a team at the client's premises. Their competitive advantage are the right processes and procedures. Because the scale of their activity is so huge, they have solutions to most known problems connected with acquiring and onboarding people into a new work environment. What's more, they usually also have access to a lot of cheap, well-educated engineers, motivated to work abroad.

Some companies always set up local branch offices nearby their clients' premises so that their staff can work at the client's site every day. They are able to provide large numbers of professional IT people in a short time. Their people learn about the client's organization and business processes for years. The clients, on the other hand, treat them like their own employees. After several years of working with so many outsourced local specialists, the customers are not able to function without them! Replacing such a supplier is obviously possible, but entails multiple risks, it's time-consuming and very costly.

It often happens that a new product supplier has more advanced domain knowledge, but is unable to provide so many local people as the companies mentioned previously. Clients mostly choose a supplier who will ensure a stable team on-site. Direct communication, everyday informal meetings and chats during coffee or lunch breaks cannot be replaced. An on-site team ensures understanding and effective information flow, which is a foundation for running projects. Customers understand it very well. That's why companies whose business model is based on the on-site presence are so successful. Although their products are not the best (sometimes even non-existent), they are able to win tenders for system integration or business process transformation.

Summary

- Forming on-site teams gives plenty of opportunities.
- Thanks to them the supplier understands the client's expectations and his organization better, makes fewer mistakes on the job, significantly improves communication.
- Investing in an on-site team enhances the relations with the customer who interprets this as a step toward a long-term partnership.
- Everyday work in the client's environment provides continuous access to leaders, allows to manage the perception of the project's progress, allows to influence the decisions made by the client, helps to protect the business against any political actions undertaken by any hostile individuals or companies.

The client may tolerate delays, but will never accept the lack of understanding of his needs or ignorance in the area of the organizational limitations. If customer's and supplier's project teams are not able to communicate effectively, the relationship between them "doesn't click," they don't fit together, the project will be stopped sooner or later. All this can be avoided if you build an on-site team.

CHAPTER

8

The Steering Committee As a Project Manager's Tool

Most people spend more time and energy going around problems than in trying to solve them.

—Henry Ford

The steering committee's meetings are usually seen as stressful events. You could almost say that for a lot of people, taking part in them generates more tension and fear than working on the project, however difficult and exhausting it is. Steering committees gather top executives and directors from the client's side as well as their counterparts from the supplier's side. It often turns out that the project team or the project manager had never had a chance to

M. Dąbrowski, *Managing IT Projects*,
https://doi.org/10.1007/978-1-4842-9243-3_8

meet these very important personas. Thus, it's natural to feel anxious about what the management's reaction will be to their presentation and how it will affect their career.

Top executives tend to categorize people and that's a fact. If they have an impression that the person speaking came unprepared, shy, or withdrawn, couldn't answer the questions in a professional and concise way, they categorize them as a poor performer. It's difficult to get rid of this sort of label. That's why the project team, especially the project manager, puts a lot of emphasis on presenting themselves in a positive light, always be right, show the outcomes of the work done, and lay the blame for all bad things on the client. On the other hand, the steering committees include the client's side, also represented by a top executive. It's important to impress them as well: demonstrate the technical expertise, show successes, minimize the negative message.

In short, you could say that preparing for a steering committee is more about the form of the presentation itself than about the actual content. People mainly focus on presenting and promoting themselves, on pretending to defend the company. In their view, a good steering committee is the one in which you don't discuss the problems, there are few comments, there are no arguments or critique, they like the ones which are quiet, cozy, and nice. The problem is, however, that such meetings are completely useless when it comes to the project. They are not constructive at all. That said, it's necessary to admit that committees are a tool for the top executives to diagnose the status, but they are the very people who expect to talk about problems. They're not interested in listening to how well everything is going. It's a waste of time for them because if something is going well, it's not worth their attention. They want to discuss the risks, solve problems, and take important decisions. That's precisely the reason for taking part in steering committees for them. What's more, the project manager should also organize steering committees to solve problems and treat them as a tool which gives him the opportunity to take advantage of the decision makers' time. It's absolutely not his goal to talk about the status. He comes to the steering committee with some proposals, and the time that he has should be used to convince everybody to accept the option that is most beneficial from his perspective.

Contrary to what many people might think, it's the project manager who fiercely fights for progress in the project that shows his best self. It's not the person who is quiet, submissive, and avoiding confrontation. The management appreciates boldness and relentlessness. From their point of view, the most valuable qualities include showing problems transparently and suggesting solutions for approval. It's not hiding uncomfortable data, avoiding stress, but the ability to hold an open discussion about difficulties and pragmatically seeking solutions. Instead of caring for one's own image, style, and the aesthetics of presentation slides, focusing on finding the guilty party, it's necessary to just focus on solving problems. It's the steering committee that is the best occasion to get the needed decisions and that's how it should be understood. It's not a meeting in which you should boast about the progress in the project, it's an opportunity to discuss some critical problems and get approval of the suggested ways to move forward.

A Pitch Deck for the Steering Committee

While preparing for the steering committee, it's crucial to bear in mind who we're going to talk to. This allows us to adjust the form of the message, the content in terms of how technical it should be, the list of problems to discuss. The prevailing mistake is to assume that the people we're going to talk to have the same level of knowledge about the project, organization or technology. Unfortunately, it's not the case, and trying to talk to them in a language they can't understand leads to frustration. For steering committees, it's the top executives. They're not interested in technical discussions and they seldom want to understand the details. They focus on the business value of the project. They focus on the outcomes, delivery dates, costs, they might be interested in how the dates affect the internal costs and the third party costs. For this reason, they expect information about the impediments which may prevent them from achieving the business goals, they want to spend most of their time during the steering committee on discussing those impediments and wondering how they can be removed.

For this reason, the pitch deck for the steering committee should be as full of content as possible while being short and concise at the same time. It should not contain more than ten slides. The status of the work should be presented only briefly, but 90% of the time should be devoted to discussions about the problems, risks, and solutions.

The typical structure of the presentation should be as follows:

- Indicators showing the deviation of the current schedule from the original one
- A very short and concise description of the progress from the last meeting and the milestones planned for the coming period
- A section devoted to risks and mitigation plans
- A section devoted to problems and planned solutions
- A section devoted to decisions to make, including the suggested options to choose from

The graphic, aesthetic aspects, or the layout of the slides are of secondary importance. You can design them according to your taste, the client's wish, or the management's expectations. It's important to present the key information in a clear and concise way, without bombarding the audience with irrelevant details which, apart from not contributing to the aim of the meeting, also obscure the picture, annoy the participants, and waste valuable time.

The status update, including the risks and problems, should be presented in such a way that it's possible to discuss them in no more than a quarter of an hour! The most important part refers to the decision that has to be made.

Admittedly, it can happen that the project is completely under control and the management's support is not necessary. We just need to make sure that we're on the right track. Most often, however, we do need some decisions because our options to act (as the project manager or the project team) are limited. So, if the steering committee takes place once a month or every other month, this time must be used perfectly. This is our "five minutes." Either we persuade the decision makers to our tactics or we get nothing and another opportunity to change the situation won't happen until the next steering committee.

It often happens that in order to make a decision the problem needs to be thoroughly understood. If this is the case, some introductory slides need to be added, but still it's a good idea to stick to the rule that the fewer slides, the better. Contrary to what we might think, preparing a short but meaningful presentation is difficult. You really have to work hard to achieve that. Managers should get only the concrete information presented in a clear and straightforward way. They have very little time.

Some examples of the presentation structure could be like this:

- One slide – status update
- One slide – risks and problems (these can also be included in the first slide)
- Four slides containing the information necessary for making the decision in which we can show the pros and cons for each of the options
- One slide to sum up the decision that has to be made

As a result, we get a presentation that consists of only seven slides, two of which relate to the status update, whereas the remaining five is a content whose aim is to support the discussion about the decision to be made. Obviously, it's just one example of proportions that we should maintain. It's also possible to have only three or four slides altogether. In most cases, it should be completely enough. It'd be problematic if we had as many as 20 or 30 slides. When it comes to presentations for the steering committee, less is more!

Controlling the Discussion

The steering committee gathers when there are some decisions to be made, that's why all the status updates should be presented as concisely as possible, without going into details. It's only an introduction to the discussion just to give some context before getting down to the conversation about the problems and risks as well as the decisions that have to be made. It's a mistake and a waste of time to discuss the progress in details, as the top executives focus on the problems and not on listening to some positive stories which are not that valuable in their eyes. The added value for the management is solving problems and making tough decisions. That's why the project manager should be able to use their time to the full.

If the project team is capable of dealing with the current problems or risks, they should briefly present the most important ones and then move on to the other issues on the agenda. Whatever is under control and does not require the management's attention, should be discussed very briefly, concisely, without wasting precious time.

From the very start of the meeting, the project manager should aim at what he wants to achieve from the steering committee, that is obtaining approval of the recommendations he had prepared.

We usually have an hour, so the status update should take a quarter, and the remaining 45 minutes should be devoted to the decision that has to be made. These decisions are the essence of the steering committees. That's why they are organized. Managers attend them only to make decisions that are impossible to make for the project manager or the project team as they are not empowered to do so.

Controlling the discussion should be a conscious choice! You should not succumb to the temptation to delve into the technical details. It seemingly shows our expertise in the area, but it's worthless for the project itself. The more time we spend presenting the project milestones in details, the less time we have left to discuss the crucial issues. Any arguments or debates which can happen also take up precious time. That's why they should be avoided altogether. You have to shortly and clearly comment on the remarks and lead the discussion toward the section in which decisions should be made. The fact that somebody criticized us, pointed out a mistake, gave us some negative feedback does not matter at all. We'll have the minutes of the meeting where facts can be documented. However, in this particular case, we're in a meeting in which we need some concrete decisions to be made.

Choosing the Right Audience

During the steering committee, we want to obtain approval for the previously prepared recommendations from the particular decision makers. Therefore, all your statements and arguments should be directed to these few individuals who have the influence and are adequately empowered in the client's organization. The project manager isn't an actor at the theater. His aim is not to get applause for his perfect presentation. He cannot try to please everybody in the room. It's impossible to present the information about the project that is satisfactory for all participants, answer all of the questions asked, address any issue, and explain all doubts. In most cases, such attempts fail. The project manager does not touch on the merit of his presentation as the discussion goes astray and the people who should be listening take over the facilitator's role. You need to know from the start who our main audience is, answer the questions asked by the decision makers, and address their concerns. How detailed the presentation is should be geared toward them in order to facilitate the dialogue and come to conclusions faster.

It's not a good idea or rather it's not even allowed to get into discussions with people who have no authority or a decision-making power!

It's wasting your team's time but also that of the people invited to the committee. It's pointless to get into discussions with the malicious people who tend to prove their higher status in the conversation, who didn't come to listen and support, but to say something pseudo-smart and show off their merits. It often seems to such people that if they humiliate the other person they will appear superior. As the saying goes: "you shouldn't argue with the stupid, because others might not notice the difference." When it comes to the steering committees, it's very accurate. Top executives often have to listen to pointless arguments which appear clownish, and the people who take part in them not only achieve nothing but also lose in the eyes of their superiors.

It's often the project managers on both sides who engage in arguments during the steering committee. What's the point of that? They can have these discussions on another occasion and there are plenty! At this particular moment, they should be talking to the decision makers and only to them. Otherwise they waste time and instead of demonstrating their professional attitude they put on a show that is annoying for the management on both sides.

You should direct your arguments only to the decision makers! They're the only ones that need to be convinced. All the other people in the meeting have no say anyway!

Summary

- The steering committee should be treated as the project manager's tool. Its aim is to get approval for the proposed tactics.
- The presentation for the steering committee should be brief and concise. Status updates should only be an introduction, the background for the main part, which is a discussion about the decisions that need to be made.
- The discussion needs to be deliberately controlled. You should not address all issues. You shouldn't waste time on irrelevant discussions. You can't afford to get provoked and engage in pointless arguments.
- The aim is to convince the decision makers to approve the tactics suggested by the project manager. This should be the focus throughout the meeting.

The steering committee is not a status reporting event by any means. If we treat it like this, we achieve nothing. Reporting for the sake of it is a waste of time. Either we want to really do something, approve a decision, discuss an issue or there's nothing to talk about and the committee at this point makes no sense. The only exception is when we need an excuse to make the executives of both sides meet so that the relations between them are strengthened.

CHAPTER

9

Arguments with the Client – The Only Thing That Matters Is Progress and Why It's Not Good to Be Right

Any fool can criticise, condemn and complain – and most fools do.

Always make the other person feel important.

—Dale Carnegie

M. Dąbrowski, *Managing IT Projects*,
https://doi.org/10.1007/978-1-4842-9243-3_9

When you watch project meetings from the side, especially those with the management and the client's representatives as participants, you can have an impression that most people speak not in order to make a valid point, but rather to show off, say something (even if it's pointless or irrelevant), perform, boast, or to blame someone trying to prove their superior position. Oftentimes futile, unconstructive arguments are initiated and their only aim is to prove who is right, which does not have anything to do with the project's progress. You rarely have the pleasure to work with people who can communicate in a professional way, have valid discussions, aren't easily provoked, and even when they know they're right, when they know that the other party is wrong, they can just keep quiet. They're aware of the fact that this is the only way to reach an agreement which will make the project team's work easier and thus the project will be finished in a shorter time.

Showing that you're right and pointing out the other party's mistakes is unwise and harmful both for the project and for the relations between individuals and the mutual cooperation of both businesses. Unfortunately, most organizations these days are based on the management concepts that promote such attitude. People are punished for making mistakes, so they try to hide them at all costs and blame others. In particular, they hide their faults from their managers who later plan work on the basis of a distorted image of the situation. The one who wins is the person not making any mistakes or fewer than their colleagues – or at least it seems so. In order to get promoted you need to be the star of the meetings, win arguments, show off in front of the management. You could think that in large and hierarchical organizations it's more often the outspoken smartie who gets promoted and not the people that actually do the work. This model is present on all levels of the organizational structure. Eventually, even the top executives try to prove themselves right when they meet with the client, when in fact it is the client who hires the supplier and expects outcomes and not proving the supplier's superior position or – what is even worse - the client's incompetence!

The preceding problem relates to all organizations, not only suppliers but also clients. The same mechanisms can be observed on both sides, which makes running the projects even more challenging. There are meetings in which fruitless arguments are being conducted, mistakes are pointed out, individuals are attacked. Such interactions don't have any added value to the project, they only antagonize teams. They promote a work environment in which uncomfortable information is hidden instead of bringing it up and having an open discussion about problems and looking for solutions. The aim of all meetings with the client is not to prove who is right. It doesn't matter at all!

The Classic Arguments

Meetings with clients and preparations for them in most cases happen according to a well-known scenario. We need to discuss the current status of the project, some problems, risks, potential delays, some generally problematic and difficult issues. Both parties prepare arguments which will support their positions. Instead of taking a proactive approach and seeking solutions, they are thinking up some defense strategies. They mostly focus on making themselves look innocent, blaming the other party, getting out of taking the responsibility. They want to look good in front of the other party but most of all in the eyes of their own management. If you listen to such a conversation from a distance, you might think that

- Nobody's made any mistake and both parties are doing their job properly.
- There's no dialogue or discussion, there's only fighting, "throwing tomatoes" at the other party.
- Neither party wants to have a real conversation.
- These are some rehearsed monologues in which the presenter publicly reads the slides prepared in advance.

Additionally, you can have an impression that the presenter's intention is not to get their message across to the other party or the partner but rather to their own management! Project managers on both sides want to show off as "pit bulls," tough players, and apt negotiators. The problem is that such meetings are completely pointless! They do not contribute to the project at all. They help no one, they do not facilitate the work, and they don't create solutions.

Running meetings in this way introduces the atmosphere of a fight. Both sides treat each other like foes. After a "successful attack," each party relishes the moment of glory, while forgetting that in this way the path for future vengeance has just been paved and mutual animosity is growing. As a consequence, it becomes a norm to point out faults, mistakes, doing things to spite the other party, mislead, and then publicly punish the other party. As you can imagine, it's not a friendly work environment, it's definitely not optimal from the team members' point of view – on both sides!

You shouldn't prepare for meetings or discussions as if it was war preparation! It's not a zero-sum game. It's not that we win and somebody else loses. It's a serious mistake. The aim of a discussion is to solve as many problems as possible and stimulate progress!

Why It's Pointless to Be Right

Proving the client wrong at all costs is a mistake. You need to think what the goal of the discussion is. Showing who made a mistake or reaching agreements which will solve the project team's problems, facilitate, and speed up work? In general, pointing out the other party's mistakes is harmful. Even if we're right, why make a fool of our partner? It's worth wondering how it will affect mutual relations. Will the partner help us in the future? Or will he rather look for an opportunity to strike back? Isn't it better to let the other person keep face? Isn't it better to focus on the objective impediments in the project, separate problems from the people and blame nobody? What advantages will it bring if we identify the guilty party? The client usually knows that something wasn't delivered, that the supplies are stuck. He also knows that the supplier knows about it.

It's especially important to let people representing the client keep their faces when the discussion involves top executives from the client's side. It will be appreciated both by individual team members and by the management. Our own colleagues will also be grateful for that. They will also make sure that we're respected in the future, they will not point out our mistakes. People will cooperate with us with pleasure. Everybody wants a partner who acts like this. What is more, top executives on the client's side will see it as a positive thing that both teams respect each other and enjoy working together. It's much more important for them than progress and performance.

Pointing out mistakes and trying to prove that you're right

- Doesn't bring any benefits
- Doesn't help, it actually obstructs making pragmatic decisions
- May offend some individuals on the client's side and even hurt their careers
- Generates personal antagonists
- Creates a climate in which people tell on each other and point out mistakes
- Creates an environment in which people act defensively, become stubborn and egoistic instead of helping out for the sake of the project (meaning both sides)
- Makes it difficult to have an open discussion based on facts
- Is a waste of time

Many people think that proving the client wrong or guilty is a manifestation of our intelligence, management skills, experience – shows how competent we are – whereas, in fact, the result is quite the opposite. In the eyes of the client, we appear as incompetent and we're just trying to conceal that by blaming the other party. Even if we are right, if the client is making mistakes, it doesn't change anything.

Our task is to complete the project for the client who is specifically paying for that. We're paid to solve problems, also the client's problems. He is spending his money on it after all. In return, he expects a proactive approach and support, not making his life difficult and showing his incompetence!

Progress – The Main Goal of the Discussion

In short, project meetings take place to make work easier, speed it up. They're about exchanging important information and delegating tasks to the people who have the right competencies. Any signs of anticipated risks or problems should be shared with team members or managers who can effectively prevent them.

Each conversation with the client should have a relevant goal! We need to know exactly what we want to achieve and focus primarily on that.

It doesn't matter if the client interrupts, insults, or points out mistakes. We need to stick to the agenda. After the meeting, the only thing that will matter to us will be whether the main goal has been achieved. Anything else is just buzz and is completely unimportant. The smart alecs should be allowed to play their game, people with inflated egos should be allowed to shine and show off, passive aggressive types should be allowed to be mean, but their remarks should be ignored. Interestingly, after several meetings, people will get used to us ignoring their malicious comments and will start to work with us based on data, objective problems, they'll just learn to cooperate.

It's worth noting that very often such behavior stems from previous project experience, cooperation with previous suppliers who weren't always proactive. They might have taken advantage of the client, they tried to make him pay extra for their services. That's why it's important to maintain a professional and proactive approach. After a while, trust will be built and the client's attitude will change and he will start to cooperate at least to some extent.

It should also be emphasized that the client's organization may force some individuals to act in an unconstructive way. Maybe it's a norm that mistakes are punished, people are humiliated. Perhaps some individuals are fighting for a superior position and each mistake made in the project deteriorates that position, makes them an object of ridicule, and provides their enemies with arguments. We need to learn to notice nuances like that. If that's the case, we should make even more effort to support our partners in the client's organization. We need to run meetings in a way which shows that we understand them, we know about their problems, their goals and despite working on the project together we want to help them. During the meetings, we can do so by avoiding criticism and attacks or pointing out their mistakes. In situations like these, it's good to sort out problems during informal meetings and agree on a common narration for formal ones.

What is a professional and proactive approach then? First of all, uncomfortable information, mistakes, or delays should be presented in a non-personal way. Problems should be separated from people who caused them. Instead of showing that an individual still hasn't delivered and blocks our work, we should say that the lack of a certain deliverable is blocking work in this area.

You need to focus on facts, numbers, and dates, not people!

Secondly, we need to focus on problems and solutions. We need to speak in technical terms, professionally, without blaming anyone. We need to show what our common goals are, what we want to achieve, what the prerequisites are for deliverables to be ready on time. Finally, we need to stop making a division between "us" – the supplier, and "them" – the client. Thinking in these terms leads to polarization between the parties and invites conflict.

There are no "us" or "them," there is one project team on both sides. Both the supplier and the client have the same goal after all!

That's why we need to be consistent in using the words like "we," for example: "What is the problem that we're facing?" and "how can we solve it?" It brings amazing results in just a short time. People stop fighting and project meetings cease to be fights and become civilized discussions which bring positive effects and really help in executing the work.

Of course, we shouldn't confuse a proactive approach with naivete or stupidity. We still need to document everything and "have everything in order." After each meeting, official minutes should be made in any form (email or a document in an agreed format). While preparing them, you should focus on the facts, numbers, and dates – without blaming or attacking anyone. You can't write that the client (or some individual) did not deliver something or that the delay was the client's fault. No intelligent person would approve such minutes. It's better to say that a given element still hasn't been delivered, which affects the delay in some specific areas in a concrete way (numbers, dates). Presenting facts in this way does not harm us! We have the contract which specifies the responsibilities of each party. We have a ton of agreements on the matter. We also have the minutes from previous meetings. All these materials let us easily find out who caused the delay. You need to act smart and clever here (in a good sense of the word). You shouldn't blame people or point out their mistakes, but diligently document the facts, while avoiding unnecessary conflict with the client at all costs in order not to ruin the already built trust and relation.

When starting the discussion, we need to think about what we want to achieve, what our goals are, think about what will help us in the project. During the meeting, we should consistently stick to the agenda, ignore mean remarks, avoid arguments which aren't constructive, and avoid quarrels. After the meeting, a note should be made to sum it up.

The aim of project meetings is not to prove who is right, but to reach agreements which will speed up the work!

Summary

- Proving that you're right or pointing out mistakes is not and can't be the goal of discussions. It only harms the relations between individuals, deteriorates the atmosphere and work environment, it creates a culture of hiding problems and doesn't allow the parties to focus on the most beneficial solutions. Also, it only shows our incompetence and poor management skills.
- The aim of every meeting is to reach agreements which will help to solve problems, mitigate risks, speed up the project, and facilitate the teams' work.
- Whenever possible, you should ignore mean remarks and avoid being dragged into a classic game along the lines of: "who's right," "it's your fault." You need to stick to the goal and consistently try to reach it.
- Any problems, risks, or delays should be presented in an impersonal way, you should focus on facts, numbers, and dates, and suggest relevant solutions, bearing in mind the project and the benefits for the client.
- You should avoid blaming individuals at all costs and let them keep their face. You should cooperate with these very people to find solutions to problems.
- In most cases, a professional attitude will be appreciated both by the colleagues and the management.
- You should always remember that the client assigns and finances the project. No sane person pays the supplier to show the client's incompetence! Instead of pointing out mistakes, we should show the client how to solve them!

CHAPTER

10

How and When to Break Bad News, or What the Client Should Know About

I'm finding that sincerity and trying to be as simple or direct as I'd like is often taken for sheer stupidity but since it is not a sincere world – it's very probable that being sincere is stupid.

—Marilyn Monroe

M. Dąbrowski, *Managing IT Projects*,
https://doi.org/10.1007/978-1-4842-9243-3_10

The line between sincerity and stupidity is very thin.

—An industry adage

Communicating with clients is an incredibly complex matter. Its form and shape are based on multiple aspects such as the client's company organizational level of maturity, the characteristics of the geographical region in which the project is conducted, ethical rules in this particular part of the world, ownership structure, that is, whether the client's organization is a private or a public entity, the industry in which the project is conducted, the flow of the decision process, etc.

A particularly interesting aspect of communication is sharing bad news about the status of the project. We're often in a difficult situation when we don't know how and when to break the bad news to the client. It may be about the delays in the production of a particular module, problems with reaching the planned performance or quality, inaccurate estimates of a specific area which apparently needs a prolonged or yet another analysis, etc. We're obviously talking here about some serious cases, not some minor delays, but about problems of great importance for the client which will affect him in a serious way, for example, when the time of delivering important business value is seriously delayed, the costs of the project far exceed the planned budget. A particularly painful consequence of such complications is, for example, the necessity to prolong the existing contracts with IT system suppliers.

Sharing this kind of news is not trivial. First, you need to be careful in crafting the message. It is important to formulate it in such a way that the client will be able to accept it. Secondly, the timing is extremely important. It is crucial to know when to do it so as not to scare the client before they get any tangible deliverables.

Situations in which it is necessary to break the bad news are objectively hard. They cause rising tension on the team. It is common that due to lack of global understanding of the nature of conducting projects or possibly little experience, project managers and team members quit, burn out, as they erroneously take pragmatism with dishonesty toward a client while total frankness would actually be harmful both for the project and for the client.

Being honest and transparent toward the client is the best policy! It creates the culture of focusing on work, facts, and solutions. It fosters a sincere and long-term relationship between companies.

In this chapter, we'll look at cases in which a project team is in a situation when they must, however unwillingly, skillfully and prudently communicate bad news for the sake of the project and the client.

When You Can and Should Be Completely Honest

Honesty and transparency should be a default policy, a dominating approach in the daily work with the client. It is obvious that over the course of the project some problems will come up. They will occur both on the supplier's and on the client's side. If the project had been properly and honestly sold, most of the difficulties that will come up will be the ones that were impossible to predict. There might be problems with team performance on both sides, or with some assumptions which turned out to be wrong. Perhaps more time needs to be spent on analyses, or it might be necessary to consult with a larger user group. There will definitely be some misunderstandings in the area of expectations. For sure, there will be some parts of the product that will require more time to be developed.

The key thing in this situation is, however, that clients are fully aware that such difficulties may occur, and therefore it makes absolutely no sense to hide them. From their point of view, problems are nothing unexpected. What counts for them is the way that suppliers inform them about the problems and how they approach the impediments. Do they act proactively or do they try to blame the client and conceal their mistakes? Do they help the client in his struggles in the project or do they just idly wait to win some time?

Transparency and honesty are relatively easy to apply in projects which were properly sold, without any hidden delays or unrealistic promises. Then there are just no circumstances which are impossible to explain rationally.

In such projects a great majority of delays can be explained rationally. There is always a cause and effect relation between the extra time needed to complete the work and the objective difficulties. You could even say that full transparency helps clients understand where the problems stem from. When we build honest and sincere relations, there aren't any problems with accepting our messages thanks to the well-deserved trust that we gained.

Even if the project had not been properly sold, which means that after the sales process, we are left with delays that cannot be rationally explained, we need to be transparent with regard to the current phase of the project.

This is essential. Clients should understand that the stages which will happen in a year or two can have realistic schedules, that when they are initiated, their estimates can be updated and the deadlines extended. It is obviously difficult to plan work that will be done in a year or two and which hasn't been thoroughly analyzed. What is true, however, is that no client will accept continuous delays in the current phase, pulling the wool over their eyes about how surprised we were with a four-week-long delay in delivering the milestone that was due in three months' time. The supplier either knew about it and hid it or they are incompetent and have no control over the project.

It is necessary to point out again that being honest, open, and transparent is the best policy. It brings a lot of benefits and gradually builds an evidence-based culture, substantive cooperation in which teams on both sides focus on the work and objective problem solving. That said, there are situations when complete honesty is harmful not only for the supplier but also for the project and therefore also for the client; sometimes some aspects of the project should be deliberately not brought up or even hidden. There are also situations when the supplier is forced to act defensively and selectively, carefully filtering the information given to the client.

Why Sometimes You Need to Choose Between Being a Liar and an Idiot

As was said in the previous chapters, the most serious problems and delays can often be traced back to the sales process, they even happen long before the signing of the contract and initiating work. They have nothing to do with performance, they are also much larger than those which you could blame the project team for. As a reminder, the possible reasons could be

- A totally unrealistic schedule which was deliberately presented to the client in order to secure the deal and meet the clients (admittedly unrealistic) expectations
- Inadequately specified or too general scope of work that turns out to be much larger than initially estimated
- A non-existent system that needs to be developed and which had been sold as almost complete or working in production

In situations like these, a project manager stands absolutely no chance to rationally explain to the client the reasons for the delays. Especially that these delays are and will seriously affect the schedule. You cannot develop a new system from scratch in three months. If the scope of work had been wrongly defined and the actual estimates are, say, twice as much, you can expect the time to be proportionately longer. You cannot tell the client that the project will be delivered two or three years later. Informing the client about it in the beginning of cooperation will most probably lead to its termination before it even had a chance to start.

The only pragmatic option in this kind of situation is "portioning" the delay. Once in a while, at the most opportune time, the supplier gradually communicates the delay, usually taking advantage of the circumstances which are helpful in finding a rational explanation. A whole arsenal of problems characteristic for this particular stage of the project is utilized. Business analysis is delayed supposedly due to the lack of dedicated staff on the client's side, the lack of legacy systems or interface documentation, another reason may be that reverse engineering of database source system schemas, that there are problems with finding the right people to conduct internal analyses. After all, it's possible to conduct multiple iterations of analyses taking advantage of the organizational chaos at the client's side when there are a lot of various user groups.

Each of the iterations brings new requirements which need to be refined for the sake of the project. Similarly, production can be delayed by claiming that there are still new requirements that must be taken into account, that the development cannot be finalized if the analysis hasn't been completed, and the needs discovered at the last moment require rewriting some key elements of code or that some refactoring is needed. You can claim that the production is complete, but internal testing is prolonged and quality is something that we do want, don't we? Another excuse might be the need to execute third-party integration tests which are often objectively unprepared, it's possible to claim that the delays in production and tests were caused by the lack of environments on the client's side and admittedly clients usually face some difficulty in effective and fast purchase of hardware.

The thing is that at some point the available battery of rational explanations runs dry. When we're dealing with delays inherited from the sales process, an inaccurate scope estimate, or a contract which promised to deliver non-existent software, the problem is much more serious. The delays in the schedule are exponentially longer. If the supplier is lucky, the delays on the client's will cover his own, but it happens really seldom. More often the client realizes that the supplier continuously postpones delivery and starts losing patience. Then every time the project manager reports a delay they have two possibilities. They can either become a liar or an idiot.

It certainly does sound ludicrous, but this is the sad truth. Project managers can say that they'd known about the delays from the start and explain why they hadn't communicated them before, which is perceived by the client as dishonesty and hiding the truth. Alternatively, they may claim that the changes in the schedule came as a complete surprise, that some problems occurred and they caused the delays. Then the supplier is seen as incompetent. The client wonders why the integrator or supplier is now and again surprised by situations which should be something familiar and predictable. Thus, we have a choice between hiding the facts and incompetence and lack of management skills.

Unfortunately, there's no simple answer as to which option is better. Both are just bad. They both entail some consequences. If the client feels that he's been notoriously lied to, at some point, he will choose another supplier. The same thing will happen if the client thinks that the supplier has lost control over the project. Everything depends on this particular client, on what he finds the most important thing at a certain point.

To some extent, every client understands that delays are common, that suppliers have a lot of projects, limited resources, varying priorities. Therefore, you could say that it's better to be a liar after all. That means: it's better to show that we have control over the project, but for a number of different reasons, we had been hiding the delays. For example, there was a tense situation with another client and we needed to move some resources to the troubled project. We assumed that we'd be able to sort out the problems on that front and make up for lost time. It's also possible to say that we had similar demands from three different companies and that resulted in the need to make schedule adjustments for all clients in order to develop the software once but in a product-oriented way.

The client cannot be lied to all the time. Incidents are manageable. Repeated lies destroys relations between the two companies, teams; they indirectly destroy the project itself and finally closes the door to further cooperation with the client in the future!

Obviously, we are talking here about situations in which the delay or poorly defined (underestimated) scope are somehow inherited from the sales process and we have a choice between admitting that the project is unrealistic, cannot be done or struggling with one of the preceding tactics. Frankly speaking, it's not very ethical, but it's beneficial for the client who can finally get what he'd bought.

Just a reminder: it's not the project team that is the source of the problem, it's the bad contract. The contract is a result of conscious decisions made by both the supplier and that client. During the sales process, both parties fool each other, which leads to some paradoxical situations: clients do this by using clever selection and negotiation techniques, suppliers – by accepting the rules of the game.

When Being Completely Honest Is Harmful for the Project

In the previous section, we talked about a situation in which the project inherits the delay from the sales process and the delay is so significant that for the sake of the project and the client, it's not a good idea to inform them about the scale of the problem right from the start as this could only lead to bringing the project to a halt before it had a chance to start properly. This situation is characteristic for very big and complex transformation projects. They usually last several years. They consist of many stages. They are dependent on a large number of legacy systems and those delivered by subcontractors. They encompass revolutionary changes in business processes, remodeling the organizational structure, which means that they profoundly affect employees and their ways of working. Even collecting requirements is complicated as the goals of particular user groups may be contradictory (the management expects changes, whereas the employees oppose them). Moreover, the outcomes of such projects are completely different from what was initially planned, but despite that, they are considered successful.

In such cases, it's absolutely impossible to prepare an exact plan for a few years. It's often difficult to define the complexity of the current stage, the next one is relatively little known, not to mention those which will be conducted in a year or two. Both for the client and the supplier, it means a lot of work. It is then advisable to focus on the current stage, and only draft an initial plan of the following one. It should all be done according to the known rule that "an elephant should be eaten one bite at a time." The entire effort is too big for both parties and it makes no sense to face it all at once. Both the client and the supplier have limited knowledge about the details of the scope for the consequent stages. Besides, their resources are not enough to deal with all of the project's areas at the same time.

On the other hand, suppliers usually try to estimate the costs of the later stages. The problem is that the actual cost of work in relation to the estimate from the sales process is higher with every day of the project when the supplier gradually discovers the client's real expectations and the complexity of the undertaking. It's usually the case that the time needed to complete the future phases of the project is significantly longer and the cost much higher. If, however, the supplier informed the client that the second and third phase will last two years longer and their cost will be adequately higher, the client may really feel petrified.

Regardless of that, there are often problems in projects, also on the client's side. The likelihood that the current phase will take significantly longer is very high. On top of that, there is the time that is needed to ensure the desired system quality. After that it is necessary to go through the procedures connected with the deployment to production, after which comes a period of stabilizing the environment. Finally, it turns out that users desperately need extra features that nobody thought about before. In this way, the current phase will take longer for objective reasons.

Such delays are characteristic for projects in which the outcome of each phase goes to production. Both parties initially think that the planned finish date of a particular phase is also the day on which it's possible to smoothly move on to the next one. Then it turns out that the deployment to production and dealing with critical business requirements results in loss of resources, especially on the client's side. As a result, the supplier gets months to deal with the backlog on their side.

It must be remembered that the client's willingness to withdraw from the project decreases with the increasing number of functionalities that are already in production. Therefore, the more phases are complete and delivered, the less likely it becomes that the client will back off from the project. That's why it's better to wait before telling the client about the delay in the next phase until the product being the outcome of the current phase has been released to commercial users.

From a pragmatic point of view, it makes no sense to immediately inform the client that the complexity of the next phase is much greater than it seemed initially, because before we even start work, the project will already be delayed anyway for objective reasons. Moreover, giving the client the news like that is harmful for him! Why run the risk of halting the project if the problems it's facing will be the same for any supplier? Why scare the client? It's in his own interest that the work is incrementally but consistently pushed forward.

On top of all this, there is also organizational imprudence, so to speak, demonstrating itself in allocating the budget in such a way that the costs rarely contribute to generating real value for the company. The main focus is boosting financial performance. In such circumstances, it's easier for the management on the client's side to deal with the gradual increase in the budget from year to year than with a sudden spike in costs that could lead to bringing the project to a standstill. Then complete honesty is also irresponsible and could harm the client.

It's important to remember that most projects are significantly delayed. With larger projects, it could even be years. If you conduct a post-mortem analysis of such projects, you can see that the supplier always informed the client of significant delays gradually, while assuring him that the following phases would be completed on time. At the same time, the work was consistently done and particular stages were deployed to production step by step. This stable delivery is key in this process. Even if the project is delayed, but incrementally delivers tangible business value, you can say that the situation is under control, at least to some extent.

The Level of Transparency Depends on the Type of Client

Mature clients value transparency and honesty in cooperation, regardless of the project progress. They are of the opinion that in order to solve a problem, you need to understand it first. Delays are not treated as something horrific. They try to analyze the possible options in order to find the best solution. Such clients prefer honest plans. Even if it turns out that the project will last much longer, they just keep their composure. They try to act honestly, they don't cheat or exploit the supplier. Most importantly, they don't hide their own problems, they remain open, and they definitely don't take the first opportunity to punish the supplier. They know that the fastest route to success is to get on with the work in an environment that fosters focusing on facts and problems. In such conditions, it's fairly easy to be open and transparent because it's not dangerous at all, such an approach is even required and expected, it's valued more than technical and organizational skills.

Less mature clients don't accept any, even the smallest, deviations from the plan. They get annoyed if there're problems. They make hasty and irrational decisions, very often based on particular people's emotions. They use every supplier's mistake as an excuse to impose a penalty or get extra features "free of charge." It is often said that they try to bilk the supplier, take as much as possible, which inevitably leads to deteriorating quality of work, as the supplier's revenue dwindles they are not able to keep the best people, which in turn results in increasing problems with carrying out the project. In such situations, the client in a way forces the supplier to act with caution. The supplier is forced to take a defensive stance which is demonstrated in filtering information that is given to the client, the supplier focuses on the client's mistakes to cover their own delays and errors. The supplier cannot tell the truth about the schedule at once so as to avoid financial trouble. The project team is then in two worlds: the internal one in which real dates and problems are explicitly discussed, and the client's world, in which it's necessary to be extra cautious in filtering and twisting the reality.

Then eventually, the project lasts longer anyway due to the fact that the objective amount of work usually requires a specific amount of time – regardless of whether we disclose the real plan or hide it and break the news of delays in small portions. In both cases, the final date of the project is the same. The difference is that in the former case, the work is done in an honest, substantive way, long-term relations are built. In the latter case, on the other hand, there's an atmosphere of discord, dishonesty, and tension; the relationship with the client is bad; and the people are mentally exhausted.

Summary

- Honesty should always be the prevalent approach. It is the foundation of trust. It creates a work environment in which the parties focus on the professional aspect of tasks, proactive way of solving problems, they don't argue about who's to blame. It builds long-term relations between particular people and companies. In difficult situations, it helps to make decisions based on the full, undistorted picture.

- With projects that have been inappropriately sold, it's often impossible to explain the delays in a rational way. Then it might be a good idea to think of gradually informing the client about them, introduce the changes into the schedule step by step, and wait for the best timing to break the bad news.
- There are clients with whom complete honesty is just not an option. It's risky for the supplier, for the project, and the client himself, as he doesn't realize the harm that he's causing with this sort of attitude.
- The willingness to back off from the project decreases in line with the business value that has already been delivered. That's why informing the client of significant delays in the very beginning of the project, when nothing has been delivered yet, is just unwise. It's also harmful for the client.
- The most pragmatic strategy is to focus on the current phase of the project, consistently make progress and deliver value to the client incrementally, without feeling overwhelmed by the problems that can occur in the following phases.

CHAPTER

11

Always Be Prepared for the Worst

Si vis pacem, para bellum.

If you want peace, prepare for war.

—A Roman proverb

The majority of coursebooks and programs on the subject of project management talks about how to conduct projects in a way that will ensure the delivery of the desired outcomes on time and within budget. The scope, quality, costs, and time – these notions have been on the minds of authors, academics, and business people alike for decades. It is generally assumed that if a project is carried out properly in every aspect, the client must be content and the project will reach its happy ending.

However, it is possible to come up with at least a few examples of projects which were brought to a halt even though they were conducted absolutely flawlessly. Most commonly, problems and delays happen on both the client's

M. Dąbrowski, *Managing IT Projects*,
https://doi.org/10.1007/978-1-4842-9243-3_11

side and the supplier's sides. Regardless of the outcomes, projects can be (and are) stopped, contracts frozen or breached. The project environment is multi-layered and complex. A project is only one part of this ecosystem. The client's business undergoes dynamic changes. Today's goals may become obsolete tomorrow. The strategy changes on the company level and on the entire group level. Executives change and that entails changes in the supplier preferences. Last but not least, there is the human aspect of things.

In one of the projects, the plan was to replace several dozen IT systems with a cutting edge platform in the SaaS model. The supplier was to manage the platform which would eventually lead to lay-offs in the client's IT department. After several years, a large portion of the platform was working on production. The client was able to reach his business goals and was content with the supplier's work. Unfortunately, at some point, the client's company was acquired by a major player on the market. The acquiring company had a completely different vision of the IT tools implementation. The result was that the platform in question was gradually disabled, which led to the project's death.

Regardless of the outcomes, almost every project can be stopped by a client for reasons dependent or not dependent on the supplier. It is important to point out that the supplier usually makes mistakes that can be used against him. Thus a project manager should always be prepared for the worst, for contract termination. Obviously, it is not the point to go to extremes, prepare very detailed project documentation, lose trust toward the client, which quickly leads to a paralysis in cooperation. The aim is for the project management people on the supplier's side to constantly bear in mind the worst-case scenario and to modify, ever so slightly, their actions; to be aware of the fact that some situations are key, to know when and how to react, how to record facts; eventually to know how to defend the company and get out of a situation in which the client threatens to terminate the contract or even to sue the supplier.

At first glance, a scenario in which the client halts the project or terminates the contract seems unreal, terrible. Unfortunately, it is not uncommon. It is obvious that neither party feels eager to share this sort of news as it is proof of failure on both sides. Contrary to what we might think, the situation unfolds according to a well-known scenario which is worth knowing and bearing in mind. It is as follows:

- Stage 1 – The client starts to make unreasonable decisions referring the project, more and more often acts irrationally, gives some signals (see in the following).
- Stage 2 – The client informs of a temporary or permanent project stop.

- Stage 3 – The supplier receives a list of demands from the client.
- Stage 4 – The supplier prepares a defense strategy.
- Stage 5 – Both sides negotiate and either reach an agreement or commence legal action.
- Stage 6 – The parties meet in court and defend their positions.

This six-stage scenario may unfold over a period of six weeks and end with reaching an agreement (stage 5) or last for months or even years, regardless of the fact whether it ends with a pretrial settlement, during the trial or with the court's verdict. From the supplier's perspective, the aim is to conduct the project in such a way that they are always prepared in order to be able to reach a settlement (stage 5), what is more, the settlement should obviously be as financially satisfying as possible.

Stage 1. Signals Suggesting a Project Halt

There are some objective reasons which should make the supplier alert that the client may act radically in the future. These are, for example:

- Recurring delays on the supplier's side
- Low quality of the deliverables
- Communication problems between the parties
- Inability to reach an agreement in regards to key issues connected with the priorities, scope, and schedule
- Evident lack of resources on the supplier's side
- Too high development costs (in relation to what was initially promised)

If the supplier has some objective problems with carrying out the project, the management should be alert to the fact that the client may lose patience and finish the cooperation.

The first symptoms of a looming catastrophe are some behaviors and decisions on the client's side which, even at first glance, seem irrational and pointless. These are among others:

- Halting the work (all the work or some parts of it) when the project seems to be under control
- Removing key people from the project (the supplier has no one to cooperate with)

- Failing to do the testing and to approve the deliverables
- No will to approve the correctly tested (by the client) products
- Delaying payment for correctly issued invoices (issued on the basis of acceptance reports)
- Blocking the work, including failure to deliver the elements necessary for continuing
- Recurring cancellation or rescheduling key meetings with decision makers

A common reason for stopping projects is preparation for selling the company or for an acquisition by another large group. The CFO's priority for the client is maximizing the company value, which means reducing all other costs (including the supplier's project). Bearing in mind the future company sales, the management has no scruples to abruptly stop all strategic investments, which become naturally irrelevant then. These are often some ruthless actions on the verge of being legal or professionally ethical. Settlements with the appropriate confidentiality clauses are signed with suppliers.

In one of the projects, a client hadn't been paying a significant amount of money for months (for a properly accepted part of the work). Next, they asked the supplier to prepare a list of redefined priorities for the project and switching the sequence of the phases, their explanation being that they needed this information to adjust to the current company strategy. The new vision of the project was presented, after which the work was terminated without a word of explanation. Within the next few weeks, the supplier received an official note informing them of stopping the project. The reason turned out to be the preparation for the sales process of the client's company. The project was qualified as a non-essential cost (investment), and as such, it was simply decreasing the company book value.

In another project, the development was stopped right after the acceptance testing phase. Despite the fact that the formal acceptance criteria were met, the client delayed signing the acceptance report. Another step was supposed to be deployment into production, including the process of migrating data to the new system. Unfortunately the client refused to send the required data package, and as a result, the work was stopped for an indeterminate period. After several weeks, the supplier was informally notified that due to some changes in business priorities on the client's side, they do not see any point in continuing the project.

It is necessary to keep in mind that the irrational behaviors on the client's side are often not incidental. On the contrary, they are planned actions, resulting from an already-made decision to abandon the project.

Stage 2. Client Informing About Stopping the Project

When the client has internally decided to stop the project, they have several options to inform the supplier. These are usually

- Formal notification
- Informal notification
- No information whatsoever

When the client feels confident, is convinced that the decision is completely justified, they usually choose the formal path (written document). At the same time, a trial may start against the supplier. The client's strong conviction that they are right may have a lot of different reasons. It's worth remembering that this is the client's point of view that the supplier may question (what is explained in the sections that follow). The client's priority is to have their decision documented for further use (potentially in court).

When the client is not certain that their actions were right, they often decide to inform the supplier informally:

- On the phone
- During a project synchro meeting at any level
- During a private meeting with the decision makers

There are also cases when the client does not inform the supplier in any form. They just break the communication, cancel meetings, and stop all the work on their side. It usually happens when the client is not sure about the future of the project, when the considered options are either restarting the project or stopping it. Most often, the reason to act like this is the client's financial situation, that is, running out of budget (for various reasons) for financing the supplier's work.

With regard to the stopping of the project and subsequent termination of the contract, there are usually two general types of these:

- Termination for a cause
- Termination for convenience

The common reasons for termination for a cause are as follows:

- Recurring delays
- Compromised quality of deliverables

- Failure to achieve the expected business results (a strongly subjective matter, but it can be regulated in a contract)
- Mismatch between the client's and the supplier's organizations, leading to continuous fights over scope definition and/or quality, what results in constant delays

A great majority of suppliers focus on preventing situations which could lead to the termination for a cause. It's also important to note that these situations are manageable.

The most difficult and the most surprising are terminations for convenience. The problem is that the supplier has no control over such situations. The reasons for this type of termination are

- Changes in the management on the client's side, which often entails different preferences when it comes to suppliers
- Ongoing sales processes of the client's company during which the CFO looks for all possible cost optimizations
- Acquisition processes (the client is planning to buy a company matching the supplier's competencies after which the project will be continued internally)
- Changes in the group strategy on an international level (individual companies, the client's company included, are ordered to work with suppliers selected for the entire group)

For obvious reasons, the client usually (but not always) will try to manipulate the facts in order to prove that the termination of the contract is for a cause to ensure achieving the highest possible financial gain! The main point of the subsequent discussions between the client and the supplier is generally to agree on the type of termination of the contract.

In one project, after a series of delays, acceptance tests were finally conducted. The client, who confirmed getting the deliverables by email, reported very high quality of the product. Unfortunately he ran out of budget for financing the further stages of the project (of which the supplier was obviously totally unaware). After conducting the tests the client refused to formally accept the product without giving a reason. As, despite the previous delays, the client knew that there were no clear arguments for terminating the contract for a cause, he informed the supplier informally that he does not want to continue the development, he won't accept the tested product neither will he pay for it. After exchanging mutual threats, the client finally decided to cover a large portion of the supplier's costs and signed a statement in which he declared his readiness for an amicable settlement of the dispute. After a while the client contacted the supplier again and asked for an offer for recommencing the project at the stage at which it was halted.

This example shows that the supplier is not defenseless, that they can and should execute their rights. It also demonstrates that you need to be assertive, but at the same time treat the client's team with respect because you can never be sure that the team isn't forced to make a decision against its will. Very often, when the business or organizational environment changes, the same people return to the supplier either to continue the abandoned project or to make a completely new business offer. It is necessary to remember that long-term business relations are more important than a one-time project and it's always a good idea to "stay in the game." It's unwise to burn bridges and fall out. The client has their constraints which must be understood.

Stage 3. Demands Sent by the Client

Generally, after several weeks after stopping the project, the client somehow states their demands. Depending on how convinced the client is that they're right, the communication can be very formal (a letter), less formal (email, status meeting), or very informal (a face-to-face meeting after which there is no "trace"). There are companies characterized by a very high level of business ethics, which treat their subcontractors in a fair way. They terminate the contracts themselves and pay for it, which means that they offer covering the costs and an adequate settlement amount. In a great majority of cases, however, the client will try to get back the maximum of the invested sum. They will not follow any ethical code but rather use any available mechanism to achieve the highest possible financial gain.

The client's strategy in this case boils down to formulating a subjective interpretation of the project, crafting a logical narration presented from his point of view. It is usually based on a selective interpretation of the facts and its aim is to lay all the blame on the supplier.

The client's modus operandi is as follows:

- To spot all the mistakes made by the supplier from the beginning of the project.
- On this basis, a logical, subjective (from the client's point of view) narration is built to explain why the project failed and the client is "forced" to terminate the contract.
- Financial demands are made.

- The client tries to enforce his conditions for stopping the project and the termination of the contract by negotiating.
- If this approach fails, the client resorts to threats of legal actions.
- In some extreme cases, the dispute really ends in court.

The message with the client's demands usually comprises two sections. In the first part, there is a description of the supplier's faults that were supposedly the reason for stopping the project. The other part states the financial demands and the conditions for terminating the cooperation.

The financial demands are usually structured in the following way:

- Penalties for delays (calculated according to the clauses in the contract)
- Internal costs incurred by the client (e.g., costs of hiring staff and subcontractors)
- Damages for lost profits (for example, the costs of legacy systems that the client was planning to switch off and wasn't able to due to the delays)

Both the issue of the supplier's faults and the calculation of the client's losses are subjective and it must always be kept in mind. They are based on a selective interpretation of the facts. The supplier can and should question them. Obviously, in order to be able to make a stand, one needs to be prepared for such a scenario from the very beginning of the project!

Stage 4. Preparing the Defense Strategy

Knowing the initial line of the client's argumentation and a list of demands, it's possible to start preparing the defense strategy whose aims will be to

- Persuade the client to recommence the project or at least show the willingness to do it, which means demonstrating goodwill to continue the cooperation (what is of significant value for the court in the future)
- Maintain good reputation on the market, which means that the contract will be terminated by mutual agreement (termination for a cause hurts the supplier's professional image in the potential sales processes)
- Limit financial losses, which means neutralizing the potential demands

- Claim damages (unsettled invoices, costs of work in progress)

The client's argumentation is subjective, it is based on selective presentation of certain facts and intentional elimination of the others. This is the foundation for the line of defense. The supplier may prepare their own (logical and convincing) project history based on events presenting him in a positive light, showing that

- The delays in the project were not the supplier's fault (or not only)
- The client is mostly (or entirely) to blame
- The contract was terminated for convenience, not for a cause

All this boils down to having the appropriate project "documentation" on the basis of which the right interpretation of what happened in the project can be made. Thus, it's important to gather this information from the very beginning of the project.

The ability to skillfully recognize which event, decision, or message is important (for the future) and then documenting it is a basis for securing the future of the project and the supplier's interests!

Apart from the so-called business aspect, it is also important to mention the formal and legal aspects that each project manager should know. In each contract, there are some clauses determining the way of accruing penalties and the procedure of terminating the contract. These issues are far from obvious and they may make it easier for the supplier to defend himself. If, for instance, it is stated in the contract that penalties for the delays can be accrued "for the exclusive fault of the client" it is then enough to prove that there was even a minor fault on the client's side and therefore neutralize the risk of penalties. Quite typical are also clauses describing the exact procedure which must be initiated in order to terminate the contract.

In one of the projects carried out in Europe, a client informed the supplier of the willingness to terminate the contract. The reason was the recurring delays. The agreed date for concluding the project was due in two months (counting from the date of the notice). The schedule of the project forced by the client showed that the project would be finished a year later than was initially planned. The client interpreted this situation in such a way that the supplier would not be able to deliver what was agreed in the contract (during the remaining two-month period) and therefore the contract can be terminated for a cause. However, the court's decision was that the supplier formally still had time to finish the project.

As it was said before, even if the project is conducted up to the highest standards, the client is happy and achieves the goals, there's still a risk that the worst will happen and the project will be brought to a halt. For this reason, the project manager should be prepared from the start, which means they should

- Document the communication with the client (important emails, protocols, meeting and conversation minutes, technical documentation, etc.)
- Record and describe any client's faults as well as the supplier's successes
- Formally follow up on any changes in the project

When it comes to the client's faults, the aim is not to point out the mistakes of the other party, which quite naturally causes defensiveness and antagonizes people. However, the situation should be described in a dry, impersonal, yet respectful and substantive manner. The project manager on the client's side will never approve a protocol which blatantly points out the mistakes made by their team. For example, instead of writing that the client represented by their employee had not made their contribution to the project, it's better to write that a specific deliverable had not been delivered, which is a blocker to continuing the work. By writing in this way, we do not attack anybody and achieve our goal at the same time.

There are often changes in the scope and priorities in the project, and as a consequence, there are some changes in the schedule. All such decisions have to be meticulously recorded in the minutes of the meetings and then included in the annexes to the contract. It often happens that both parties agree on some changes in the schedule, start the execution phase, and forget about signing the annexes. Then there are some formal problems with issuing invoices, and when both parties decide to part ways, it's difficult to decide which plan is formally binding. For the supplier, a situation like this is very harmful. They have to prove in court that their version of the facts is true. For the jury, it's always just a subjective opinion of one of the parties. It turns out that the client often claims in court that they had not accepted the changes in the schedule. On the other hand, if all changes in the contract had been formally confirmed and included in the contract, the situation becomes evident and objective.

Most people tend to ignore the habit of documenting things. For them it's a waste of time, they deliver on time, there are no problems, nothing bad is going to happen, then why bother with unimportant details – there's no time for formalities, it's necessary to focus on the important stuff and on the work to do. It's a very serious mistake!

Facts which hadn't been documented don't exist in any legal proceedings. If there's no trace of some event in the project, it's as if it had never happened!

Even if the project status is really good, the project manager should generate documentation anyway. He can write an email to the client's management and describe how the project team (on the client's side) appreciates the progress, the quality of the deliverables, or the product features. The good condition of the project should be emphasized in the meeting minutes. Why is it so important? Because when problems arise at some point (and they sure will), it'll turn out that such documentation will be the supplier's greatest treasure, the foundation on which they will be able to build their defense strategy. Therefore, formal protocols, meeting minutes, and the right standards of storing and categorizing documents are absolutely not a waste of time. Quite the opposite: they are the project manager's basic and essential duties.

A classic example from the IT industry begins with signing a contract based on a high-level, very general description of the scope. During the project, both parties tend to interpret the contract in a way that benefits them. It's obvious that the supplier tries to simplify the work, focus more on the product, and limit the costs on their side. Whereas the client tries to prove that the scope was bigger, that all requirements have to be met, even those that are not mentioned in the contract, as it's "obvious" that their implementation should be compatible with the client's point of view and needs. Such situations often lead to delays supposedly due to the supplier's fault, when in reality, the reason for them is the difference in the interpretation of the scope, the supplier's cost continuously grow, which in turn leads to extending the time, to financial losses, and often to ending the cooperation.

If such a case goes to court, the client's strategy is to blame the supplier for the delays. The supplier's line of defense focuses on proving that regardless of the signed contract the client consistently forced them to perform additional tasks which were not mentioned in the contract. In order to be able to apply such a strategy, the supplier must have the right documentation, which will support this version of the project history.

Stage 5. Negotiations

When both parties have flung accusations and demands at each other, when the emotions have abated, there's time for a pragmatic assessment of the possible options. Both the client and the supplier make an attempt at proving the other party's fault are stronger and their demands more justified. The goal here is to strengthen one's bargaining power. It's often the case that one of the parties is really worse off. However, usually if the supplier had gathered the proper documentation of the project, it's possible for them to question and neutralize the client's argumentation, to prove that the situation is not as obvious or clear, and that suing the supplier may end badly for the client, which suggests that an amicable decision of closing the contract and reaching an agreement is the most beneficial resolution for both parties.

The client analyzes the following aspects of the situation:

- How likely it is that they will be able to prove the supplier's fault
- How high is the claimed amount that they can potentially win in court (penalty for delays, lost profits, staff costs)
- How long it will take and how costly the trial will be

The supplier needs to assess

- How strong their arguments are along with the subjective history of the project that might potentially be proved to the judge
- What mechanisms of executing the claim amount are employed in the contract
- What the costs and time of the trial will be

Once again everything boils down to assessing (in regards to the future trial) the strength of the arguments on the basis of which both parties build their subjective description of the project and calculate the amount of claimed damages. That is exactly why the winner is the one who is better prepared and who's done most work, the one who has better documented facts; the one who took great care to keep the documentation in order; the one who will spend more time on a thorough and substantive preparation work before the negotiations.

In the example cited earlier, the client stopped the project and at the same time refused to settle an invoice for a substantial amount. The supplier was well-prepared for defense: within several weeks of the negotiations they presented a thoroughly documented project history after which they threatened to take the case to court. The client calculated the legal costs as almost twofold higher than the amount that the supplier rightfully demanded in the first place. As a result, the client finally decided to pay most of the invoiced amount and ending the project by mutual agreement.

In the course of the negotiations, the client shouldn't be treated as an enemy, the conflict shouldn't be approached personally or emotionally. It's important to try to understand the client's constraints (as there are surely some and their actions do not stem from sheer malice). It's worth looking for a pragmatic solution, not the one that we see as "fair" or the one that "should be reached," but rather such that is possible to implement.

A common scenario for agreements is the so-called zero option meaning that demands are dropped and a settlement is signed. Each party must then reconcile with the accrued costs. It is the case when the situation is so complicated and arguments on both sides so strong that nobody wants to take the risk of initiating a trial.

The supplier should be aware that the stopped project is just one of potential areas for cooperation with the client. If the works are stopped for reasons objectively beyond the supplier's control, it's not a good idea to burn bridges and ruin the relations. A good option is to balance the client's demands with potential discounts for future orders. At the time, these are some virtual costs for the supplier, but they motivate the client to place orders in other business areas. They also show that the supplier thinks constructively and is a good partner for the future.

Stage 6. Legal Proceedings

Clients often threaten to execute their rights in court, as it's their privilege, so the supplier should meekly grant all wishes, even though they hadn't been included in the contract. If the project is conducted diligently and properly documented and objectively there are no major faults on the supplier's side, then despite the circumstances they can feel relatively safe. Obviously, legal proceedings are something that should be avoided at all costs for a variety of reasons. It's a waste of time, reputation, and money for both sides of the dispute. It's absolutely not true that if a project is delayed the client can simply sue the supplier and claim damages. The court trials can take years and their results can be effectively shaped by building a defense strategy based on strong arguments.

The legal proceedings usually follow a specific process:

- First, the court inspects the formal issues, for example, if the notice was submitted properly (often it turns out that it wasn't!).
- It's also verified whether the parties had tried to reach an agreement before they initiated the process (that's why the supplier should take several attempts at that and document them properly).

- Regardless of the preceding, the court (obviously lacking the necessary expertise in the field) appoints an expert whose role is to trace the "objective" course of the project (which usually takes years) and finding out how each party contributed to the situation at hand.
- The judge's verdict is often based on the expert's opinion.

Therefore, the expert's opinion is crucial. This person is usually an independent specialist who tries to objectively analyze the project. They work in the following way:

- Asks each party to prepare a description of the project history and all data supporting the "subjective version" of what had happened seen from both perspectives.
- Analyzes both versions, checks whether the data is consistent with the description, asks some follow-up questions, holds meetings, etc.
- Finally, issues a preliminary opinion and sends it to both parties, the opinion includes all the inconsistencies that have been identified.
- The parties have a possibility to send (in iterations) letters with some comments and explanations (which can last for years).
- When the expert sees that further exchange of letters is pointless, they issue a final opinion and submit it to court.

Thus, it can be noticed that the key to winning in court (apart from hiring the best lawyers) boils down to building a convincing history of the project, supported by proper project documentation.

It's worth noting that the project documentation are not only protocols but also email communications, presentation slides, minutes of the meetings. A smart project manager can gather all important facts for the future at a minimum effort.

Summary

- A project can be brought to a halt at any moment, regardless of the supplier's work results.
- Negotiation of a settlement and the court trial are based on the subjective argumentation of both parties, which in turn is based on a properly prepared project documentation.
- Non-documented events might as well be treated as ones that never took place.

It must be emphasized that the aim is not to make the project manager drown in tons of papers and be too busy for his proper work. In case of doubt about what is important, what to record, it's a good idea to ask yourself a question: "Would that information be relevant for a judge if the client sued us today?"

CHAPTER

12

Penalties – Much Ado About Nothing

Things get worse under pressure.

—Murphy's law

Most contracts in the IT projects are written on the basis of the "stick and carrot" rule. Thus, we have some benefits in the form of revenue that can be invoiced after a certain stage, phase, or milestone of the project has been delivered and formally approved. There are often some other bonuses such as a success fee paid if the project is finished exactly on time. On the other side of the spectrum, there are penalties calculated proportionately to the length of the delay. Clients are usually worried about losing control over the work schedule on the supplier's side. That's the reason why, during the sales process, they enforce including a mechanism that, by threatening to reduce the financial gain from the project, will make the supplier alert throughout the project.

M. Dąbrowski, *Managing IT Projects*,
https://doi.org/10.1007/978-1-4842-9243-3_12

As a consequence, during the project, both the management and the project team should always bear in mind that apart from managing the delays, solving problems, and removing impediments, they must be aware of the threat of penalties. Therefore, they must act carefully, document project events, and effectively protect the company's interests by neutralizing any attempt at charging penalties.

It is sometimes the case that clients have virtually no choice and cannot influence the supplier in any other way, so they resort to threatening to charge penalties on the delays. It's not an ideal technique for improving cooperation between the companies, but it's fair to say that sometimes it's the only way to motivate the other party. Of course, there are clients who just like to manage suppliers in this way and charge (or at least try to charge) penalties automatically for every minor fault or for missing the deadline.

Management reacts very nervously to any threat of penalties. The thing is that it's not that simple. It does not happen automatically. If the contract is properly formulated and the project is conducted with caution, the project manager is able to effectively neutralize most attempts at charging penalties. Clients almost always have problems with justifying the amounts and providing clear evidence for charging penalties. Not to mention the fact that it's absolutely a last resort, and after penalties have been charged, it's impossible to improve the relations between the companies, which leaves them only with some legal actions.

In this chapter, we'll explain how and why penalties are charged, and why it's pointless and harmful for the client. You'll also learn how to protect yourself against penalties. We'll show you that even if the penalties are formally charged, we still have a few options, in fact, we can even turn the situation around and persuade the client to extend the cooperation with this particular company.

How and Why Are Penalties Charged

Clients worry that they'll lose control of the project. They know that the supplier works for various customers. There's always a risk that some other company will offer a better deal and pay more. It means that the supplier's priorities may change at some stage of a project. That it will happen is usually close to certain, and thus, as early as the sales stage, while preparing the contract, clients try to ensure that it will contain a mechanism for seemingly controlling the supplier, increasing his motivation, and finally bringing back the right priority of the project done for them. This mechanism are the penalties discussed previously.

Clients think (and wrongly so) that without those controlling and oppressive mechanisms, the supplier could extend the schedule as he pleases. It must be remembered though, that projects are conducted for purely economic reasons. Therefore, the supplier's main motivation is to issue invoices as early as possible, while generating as low costs as possible, which basically means delivering on time or even sooner. Of course, companies usually give priorities to their clients on the basis of the business development potential and the volume of expected revenue. The more a client pays, the more serious treatment he gets. Such a client also gets better specialists and gets his products sooner. That's why penalties can be useful. If, for some reason, the supplier's motivation level drops, as other customers get his attention, the client may or should resort to charging penalties.

On the other hand, if the supplier does his best to make up for the delays, tries hard, and is motivated, charging penalties is completely pointless, does not bring anything positive to the cooperation, ruins the relations and discourages from further cooperation. It's also important to check why the supplier's motivation dwindled. It's often the case that the revenue is too little in relation to the scope of work, which means that there were wrong assumptions in the contract. It may turn out that the client interprets the scope of work very subjectively, constantly forces the team to do extra work, which results in higher costs on the supplier's side and that in turn leads to doubts that the project will ever be profitable. Then it becomes obvious that the supplier's motivation is lost and people will be moved to some other, more profitable projects. **In such cases, even the threat of penalties is not enough and clients should consider extra financing and paying for the additional (from the supplier's perspective) work.**

Unfortunately, due to some sort of peculiarity of most organizational structures, their managers focus more on protecting their own positions rather than on the actions with the aim of gaining objective benefits for the company. It's easier and safer to say that the supplier isn't delivering (while overlooking the cause of the problem), that penalties should be charged. It's more difficult to admit that there are new requirements in the project, the costs have increased and for the supplier to avoid losses or preferably to make a profit (and want to deliver on time), it's necessary to pay more.

How does the penalty mechanism work then and when is it created? It's defined over negotiations taking place during the stage of drafting the contract, still in the sales process. Normally, there are penalties for delays in finishing some specific stages of the project or milestones. They are not charged straight away. The supplier is given a certain buffer, the time in which despite delays in the schedule no penalties are being charged. This is a grace period which usually lasts from two to four weeks. After that penalties are calculated proportionately to the length of the delay in completing a specific milestone. Depending on the contract, the penalty amount may increase with each business day, calendar day, or a week. The amount by which the penalty rises in a unit of time is also a matter discussed in the course of contract negotiations. It's usually a fixed amount or a percentage of the value of the specific milestone or the project. It is also possible that the whole penalty for a delay in delivering this specific milestone (or phase) may be limited to a certain amount, which usually means a percentage of the value either of the delayed milestone or the whole project. An example mechanism of penalty charging could be as follows:

- 0.1% of the value of the whole project for each business day of the delayed stage.
- Charging penalties starts if the delay exceeds two weeks.
- The entire penalty for delaying a certain stage can't exceed 30% of its value.

It is important, if not the most important, not what the exact values are, but rather the provisions specifying the rules for charging penalties, determining when the client can start charging them. Thus, there are the following scenarios:

- The penalty is charged when the project is delayed.
- The penalty is charged when the delay in the project is the supplier's fault.
- The penalty is charged when the delay in the project is caused solely by the supplier.

The first option is the worst one because the supplier takes the risk of financial repercussions regardless of the work outcomes. The supplier is not fully in control of the situation. It's enough that the client does not meet the deadlines on his side by which the supplier's work will be blocked or delayed and according to the contract penalties can be charged anyway. The second option is a little better. The client can charge penalties only when it's possible to prove that the delays caused by the supplier affect the project schedule.

From the supplier's point of view, the last option is the most comfortable one as the penalties will be charged only when the delays are caused solely by the supplier. The word "solely" combined with the correct running of the project and properly documenting it, almost fully protects the supplier who will always be able to prove that the encountered problems were somehow linked to the fact that the client did not do everything they were supposed to do. In order to prevent charging penalties, it's enough to show any reasonable cause and effect relationship between the client's and one's own work.

It should be emphasized again that the details of charging penalties are the subject of negotiations between both parties and are prepared at the stage of drafting the contract. It's the best time to protect the project from penalty clauses! It's especially important to ensure that any penalties will be charged for delays caused solely by the supplier. Only then should specific values be discussed.

Why Charging Penalties Makes No Sense

In a nutshell: the threat of penalties is only effective until they're formally charged. When it happens, the pressure disappears. A potential risk becomes a fact, a real problem. Not only does the supplier lose motivation to try to speed up the work and make up for lost time, he also faces the problem of deteriorating the financial situation of the project and from then on focuses on protecting his interests. He switches from being proactive to being defensive.

Therefore, what potentially works is the threat of penalties. Clients often calculate penalties and show the amounts to the supplier as a scare tactic. The aim is to exert the right pressure and it does motivate both the project team and the company management on the supplier's side. It sometimes happens that for various reasons the resources had been moved to other projects by the supplier. It may result from all sorts of problems reported by other clients or from trying to optimize the invoicing schedule to obtain the most optimal annual financial results. Then the threat to charge penalties makes sense, may be utterly effective.

You should bear in mind that the mechanism works only when the supplier can act, meaning that more people can be assigned to the project, people can be motivated, priorities changed – generally speaking, work can be done faster. If the supplier is already doing everything possible, tries hard, treats the project seriously, but is facing some objective problems, threats of penalties are good for nothing! What would the expected outcome be? What value added to the project or for the client could it bring? In such cases, threats of penalties may only put unhealthy pressure on the team and deteriorate the relations between the two companies.

For this reason, clients very seldom charge penalties. If they do, it's because of despair and helplessness, but it's also a sign that they're planning to terminate the contract, so basically they don't care anymore. **From a formal perspective, charging penalties practically puts an end to the cooperation and that's why it's pointless.** If the supplier is charged a high (and not symbolic) penalty fee, he will get rigid and start to act with extreme caution. He will stop investing in the project, will withdraw the best people, and will switch into a passive mode, he will start collecting evidence against the client and then wait for the client's mistake, which will be used mercilessly, etc. It will quickly lead to deteriorating the quality of the cooperation and ruin the relations and work environment. It should also be remembered that all this is happening when the delay is the reason for the supplier's increasing costs and decreased invoicing (spread over a longer period due to the delay). Charging penalties only adds insult to injury, that is, it negatively affects the financial results of the project even more, which means that the client's position gets worse and worse from the supplier's perspective. Eventually, the client gets an effect which is completely opposite to what he wanted, instead of motivating the supplier, he discourages the other party. Instead of speeding up the project, he slows it down.

Consequently, charging penalties is harmful even for the client himself. In fact, it almost always leads to terminating the contract and to legal actions. And indeed then it's the only case when applying penalties makes sense: during the trial, the client will have time and conditions to claim compensation for the costs of delays.

Only charging a symbolic amount (rather insignificant for the project financial status) may make some sense, it may show that the client is desperate and cannot accept any more delays. But even in those instances, the supplier will start to act less flexibly, which is not beneficial for the client.

How to Protect Yourself Against Penalties

The main protective mechanism against charging penalties is a properly negotiated contract. It's during the sales process that certain provisions (making it impossible to persecute the supplier) can be introduced to the contract. The most important defense measure are the clauses referring to the conditions under which the client can start charging penalties. **A reminder: it should be written in the contract that the penalties will be charged only when the delays are caused solely by the supplier.** The word "solely" is extremely important. It paves the way to a lot of defense tactics. If such clauses cannot be negotiated, charging penalties for delays caused by the supplier is also manageable. What is absolutely unacceptable is a provision stating that any delay in the project (without specifying the guilty party) will result in charging penalties. In this case, the supplier will have absolutely no chance for effective defense.

Thanks to the provisions mentioned previously, the supplier may easily defend himself. It goes like this: it's important to meticulously gather project documentation, with a particular focus on monitoring any, even seemingly minor, deliveries from the client's side, in order to formulate a subjective interpretation of the delay, such as in which the client caused the delays on the supplier's side, which made it impossible to finish the work on time. It's clear that if the clause stating "caused solely by the supplier" is in the contract, then it's substantively impossible to charge penalties. Projects are complex and it's nearly always possible to prove that the delays were somehow caused by the client.

Of course, it's only possible when the project is conducted thoroughly and meticulously. **Thus, a key to protecting oneself from the penalties is documenting all important events in the project, on each level and**

in every area! It's important to monitor the work on the client's side, oversee the deliveries and their impact on our own deliverables. It's a good idea to collect any analytical documents, prerequisites for migration work, descriptions of database patterns of the source systems, data samples, interface specifications, etc.

Contrary to what most people think, it's not so easy to charge penalties if the project is properly and skillfully conducted. On the other hand, the problem cannot be ignored altogether. If the client is talking about charging penalties, it's a sign that the situation is getting serious and the client is considering different options, not necessarily beneficial for us.

It's worth remembering that projects are conducted to make a profit. It means that it's necessary to issue invoices, the sooner, the better. So, it's also good for the supplier when the project runs on time. **It's no reason for joy when there are delays on the client's side, there's always a risk that the client will stop the project irrespective of the outcomes on the supplier's side**. For example, the client might come to the conclusion that he's not on the right track and the whole undertaking makes no sense.

There's some sort of dualism: on the one hand, it's necessary to deliver as early as possible, solve all problems regardless of who is to blame, always help the client. On the other hand, it's necessary to constantly gather documented "proofs" in order to always be ready for the worst to come, so have the evidence to prove one's own version of the course of events to prevent charging penalties or to protect oneself in court.

How Clients Use the Penalty Clauses

As I said before, a threat of a penalty fee is a motivating factor. The fear of punishment brings positive effects, but formally charging them leads to irreversible harm to the project and to mutual relations. **For this reason, clients usually estimate potential penalties, but they don't actually charge them.** They use them as a scare tactic, and more often they just try to negotiate some extra benefits. For instance, they offer trading the penalties for some useful functionalities, licenses, or free services that the supplier can produce or provide free of charge in return for balancing the penalty fees and withdrawing demands. Clients usually want

- Licenses for extra users
- Licenses for modules that they had been initially planning to buy
- Functionalities included on the roadmap of the developed product

- Identified changes in the scope of the project, change requests that they would have to pay for otherwise
- Extra trainings
- Extended maintenance period (after the implementation)

Therefore, they are meticulously calculating the amount of potential penalties in order to use them as a bargaining argument and get better buying conditions for the services that they would have to buy in the future anyway. Both parties are fully aware of that from the very beginning of the project.

For the supplier, the option is to neutralize the penalties by trading them for discounts on offers for completely new products.

Thanks to that, revenues from the current project are not lowered, the problem is postponed until an undetermined moment in the future. What's more, future projects are yet to be estimated so there will be some space to act, a lot of possibilities to blur (and neutralize) the discount. An important asset of this solution is increasing the chance to get a new contract with the current client who is getting a discount for products and services offered by the supplier in the future.

Summary

- The threat of penalties works only until they're formally charged.
- Charging penalties makes no sense, because it is practically the same as stopping the project and suing the supplier.
- If the project is taking longer for financial reasons, the client should consider investing more money, as it will help the supplier get more resources and speed up the process. Charging penalties in such situations brings a completely opposite effect, making the partner less flexible and causing further delays.
- In most cases, clients are interested in trading the potential penalties for free functionalities, services, licenses, or discounts on future offers.

CHAPTER

13

Approaches to Project Management – Academic Discussions vs. Real Life

In theory, theory and practice are the same. In practice, they are not.

—Albert Einstein

M. Dąbrowski, *Managing IT Projects*,
https://doi.org/10.1007/978-1-4842-9243-3_13

It is a common belief that a management methodology or method is a magic recipe for most problems that can come up during the project. Thus, we have various standards, recommendations, certification organizations such as the PMI, PRINCE2, Scrum, SAFe, etc. Some of them are considered complicated and obsolete, whereas others are thought to be trendy, more effective, and modern. There are entire businesses created around them; their activity includes inventing new versions of recommendations or standards, sales of educational materials, certifications, creating consulting positions, or company transformations. All this contributes to a certain type of trend, a pressure to apply these particular management techniques. As a result, most of the people beginning their project management career think that obtaining a certificate in a certain method or standard is a proof that you have the skills necessary to manage complex projects and is a formal piece of evidence about your experience and even guarantees success. It is also commonly believed that if a project failed or is delayed, the reason for that must be that the "holy principles" of a given management technique had not been adhered to.

It must be said that it is not the aim of this chapter to criticize any project management standards or methods. They're all based on reasonable premises. In most cases, it was years of accumulating experience by a lot of companies and in various different industries which led to their creation. The thing is that each of them differs from the others, has some pros and cons. You can't blindly apply them to any project. It's not true that agile methods solve all the world problems. It's not true either that obtaining certificates is the same as having experience. A certificate confirms the knowledge of educational materials prepared for this particular exam. It's running projects that gives you the experience in project management! Only after several failures and successes can you accumulate the practical knowledge, begin to understand certain issues, in particular those related to the characteristics of a supplier's or client's organization or the so-called soft skills.

The choice of a management method cannot be dictated by the current trends, buzz around the method, the fact that we would like to learn it or a freshly earned certificate. It is also not a good idea to implement any method blindly just because something is stated in the books or educational materials. You work in a specific company, for a specific client, you will be implementing a specific product in a specific industry, work with specific people who have specific skills and experience. All these elements make the project unique and it has to be taken into account while selecting the method for running this particular project. The popular methods should be seen as a collection of some good practices which are indeed effective in most cases; it's a good idea to learn from them, but you need to be reasonable and make informed decisions about the aspects that will need to be adapted to our specific needs.

The Most Common Mistakes and Misunderstandings

In the following sections, we are going to analyze several examples showing the consequences of a forceful implementation of some methods, a lack of its understanding or ignoring the wider context and the environment in which the project is executed.

Example 1. Treating agile methods as an equivalent to acting without a schedule or time pressure and allowing for delays

When agile approaches, such as Scrum, are applied, there is often a clash between two realities: the client's and the management's vision and the perception of the development teams. The former ones have specific expectations when it comes to deadlines and they even assume that the selected method will ensure that the work will be done faster. They operate on specific company objectives that must be achieved in a specific time. On the other hand, the supplier often has to deal with problems related to how the people at the company who see agile methods as those in which they are the ones who tell the client what and when they will manage to deliver and how much it will cost. They think that there's no time pressure in the project – in a nutshell, they see agile as "a lack of a fixed schedule and time pressure." It's a huge misunderstanding.

Agile methods are not about a lack of a schedule, uncontrolled implementation of the client's wishes, or a lack of any cost control procedures. There's always a high-level schedule, mostly because projects are executed in order to achieve business goals which are always time-related. In this particular case, an agile method should ensure a shorter feedback loop so that the decisions concerning features, dates, problems, or resources can be easily adjusted to the current situation. That's why they're called *agile*.

Example 2. A hybrid project in which an "open bar" approach (flexible scope) is combined with an expectation of a fixed price.

Another example of an incorrect implementation of a project management method are projects in which clients expect an agile approach mostly because they see it as an opportunity to be able to add unlimited functional requirements throughout the project without any respect for the supplier's production cycle. Interestingly, they also assume that the project's budget and schedule will remain the same. They're annoyed when the supplier escalates the impact of the bigger scope on extending the deadlines and requires a higher budget. To sum up, the client's understanding is that he's buying a complete project with an (almost) unlimited scope, but with a fixed date and cost.

This expectation seems contradictory in itself; in practice, the situation is hardly ever that clear. You must bear in mind that it's extremely difficult to achieve all these three goals at the same time. You can limit the budget, but then the scope has to be flexible, which may lead to a situation in which the product will have fewer features than initially expected. You may as well expect all the features to be delivered, but then the cost may increase. The problem is that if the supplier gets a list of requirements and the budget that must not be exceeded, then at some point, he will have to stop taking client's feedback on the features into consideration. Otherwise, he will start losing money on the project.

In other words, the agile approach will function well at the start, but will be gradually becoming pointless when the budget limitations are reached. This problem can be solved in two ways. It's best to reach a financial agreement so that they don't stand in the project's way. It can be assumed that the most important parameter is the fixed budget, but then the client should bear in mind that he will have to keep his expectations under control throughout the project.

Example 3. An attempt to forcefully adjust the production cycle to two-week iterations

An interesting example to analyze is a recurring pattern of the client's and/or management's unwavering attempts to adjust the internal software development cycle to a chosen project management methodology. Most often we see it with Scrum, which is an expectation of iterative and incremental delivery every two weeks. And in most cases, the transformation is indeed possible. However, there are situations where some modules, usually involving complex business logic or requiring a more traditional approach, call for longer release cycles, lasting, for example, three months. Forcing the team to present usable demo versions every two weeks (as is prescribed by Scrum) is counterproductive or even ineffective and delays work in the long run.

A good example to demonstrate the preceding problem is a large project run for a European company whose estimated scope was more or less several dozen thousand man-days. The client pressed for applying Scrum from the very beginning. External consultants had been hired and their role was to help in implementing the method at the supplier's organization. After some time, it turned out that when it comes to some modules, working in two-week release cycles makes no sense because the domain characteristics call for a traditional waterfall approach consisting of an analysis, making a technical design, implementation, and tests conducted at the very end, which meant that the incremental delivery would be possible every three months. Eventually, both sides created a project management methodology which was a hybrid of various approaches, but it worked well, was effective and efficient.

Example 4. No alignment between the chosen project management methodology and the organizational culture and the company's way of working

A frequent mistake is an attempt to apply a project management methodology which is completely incompatible with the company's way of working, either on the supplier's or on the client's side. When we suggest an agile approach and the client's organization is a large, hierarchical corporation where the decision processes are multi-layered, take a long time, competencies are sliced among mutually hostile siloes, where the entire company is based on formal project management rules, we are dealing with a serious problem. If the client's organization is new for us, there's little chance that we will be able to understand it instantly and decide on the best project management methodology.

Nowadays, many companies decide to apply agile methods. More and more often, these are large enterprises, huge organizations with thousands of employees. The management instinctively senses the benefits of being agile, but at the same time, plays down the cost and effort needed to transform old and stiff ways of working into a new approach. While conducting a project for one of the largest organizations in Europe, a decision was made to implement agile, but the transformation was not entirely successful. The main reason was that it was impossible to have people readily available for various different types of tasks, be it business analysis or collecting feedback on the system. Each problem that was to be discussed had to be announced in advance in order to check and secure availability of the needed people, wait for weeks on end. At the meeting, the client's representative expected the supplier to have pre-prepared questions and answers. What was needed was a dynamic exchange, brainstorming, gathering data, whereas each follow-up meeting with a particular person had to be planned a week or two ahead! The management wanted to run an agile project with only high-level descriptions of requirements, but the organization was completely unprepared for that. It just worked in a different way. The main problem was that it was unable to build a cross-functional, committed team. Its way of functioning was based on a rigid management structure, in which all decisions were made by line or department managers. Even preparing a simple demo of the system and gathering feedback amounted to a logistic challenge.

Example 5. Sticking rigidly to a selected project management methodology, even if the work can be done better, faster, and more effectively

In case of waterfall-type method, the problem is manifested in the form of delaying the development phase until the analysis gets the final approval, when some areas have already been defined and work could start, waiting with testing of the working functionalities for the whole package to be complete, or delaying feedback until the time of delivery of the complete system.

In case of agile methods, the main mistake is sticking to the two-week-long release cycles in all areas, even when it doesn't make much sense. For complex modules, where the development time frame for any usable features is two or three months, preparing demo versions every two weeks is pointless, because there is nothing to show and also extra time is needed to release a version, prepare the installation, etc. In events like these, it's more effective to wait until we have something to present.

How to Select the Right Project Management Methodology for Your Project

Following a single project management method is usually a mistake. You shouldn't decide on a specific management approach just because you're familiar with it, because it's trendy or because you want to gain experience. There's no perfect methodology, which is always best and solves all problems. Each of them is flawed to some extent. The decision about which approach is best suited to your project's situation belongs to you. What you also might need to take into account is which aspects of specific methods to use in order to increase the chances of delivering the project on time.

You should be neither ashamed nor afraid to invent new solutions! The fact that something is not a standard approach, or it's not applied in other companies, does not mean that it will not work in your situation. And vice versa: the fact that some approach is effective somewhere else doesn't mean that it will work for you.

It's not obligatory to choose one specific or trendy project management methodology. They should only be treated as a set of good practices which are worth analyzing and then think carefully about what to use and how to apply it in your project.

It might happen that you decide to implement a given methodology and follow all the steps and guidelines from a book or manual to the letter. You might hire an external company to help you organize that. You might also want to fall back on extensive experience of a particular company, in a particular industry, with a particular type of product and for a particular type of client – perhaps your company will succeed thanks to introducing an innovative method. Someone might say that "your methodology" is just a combination of several known standards or guidelines, that this is not how things should be done, that it's not recommended. But does it matter? No! if your project management approach, your methodology works, it's the best proof that you're acting in a proper way.

Besides, most methods or standards need to be adapted to a particular company or industry anyway. It's actually a very lucrative business. You must be aware that any project management standard or methodology is also about making a profit. Why are new versions published? Why do you think certificates need to be constantly renewed? What are those key changes happening over a period of two years that call for a new version of a given methodology? Somebody just invents them, somebody writes books and sells them, somebody holds exams, there are also websites, blog posts, seminars, debates about the primacy of one project management method over the other.

You need to remember that a methodology must be adjusted to a specific task at hand, to the context in which it needs to be solved. For example, agile methods are trendy, but can they be fully applied while working to invent a new vaccine or a new plane model? To some extent, maybe they can, but at a higher level, we need phases and sequence and some high-level budget limitations. Who said that the rules of PRINCE2 or PMI can't be combined with agile methods. What is more, it even makes sense. There are some discussions about project management methodologies and methods where PMI is the dominating methodology, but Scrum is a framework used for a specific implementation.

It's important to remember that methodologies and methods are just a collection of some guiding principles and not secret recipes for success! They must be treated with some reserve and approached critically (critically, which doesn't mean negatively). You shouldn't hastily select a project management approach and implement it blindly. Then you often put the organization in a limbo and unnecessarily prolong the project by implementing mechanisms which are an obstacle to people's work and are totally incompatible with a given domain or product to be developed.

What usually works well is inventing some sort of a hybrid methodology which should encompass the following aspects:

- The so-called DNA of the supplier, which means some specific processes and experience, the organizational structure and escalation path, embedded production cycles
- Project management methodology expected by the client
- Project management methodology currently used at the supplier's company
- Project manager's working style, his experience, and competencies
- The current state of the project organization – how many people are available, how competent they are, how many seniors we have on board, which translates into how we can delegate responsibility
- The client's working style and organizational structure, whether the client will be able to ensure a team and particular people's time necessary to work in a given approach (also the approach that is preferred by the client)

- The current state of the product to be implemented – whether it's created from scratch, it's possible to act dynamically and prototype, or whether it's an old legacy system implemented at many client's companies and introducing any changes is slow, requires synchronizations between various installations, the code is complex and introducing changes is a tedious process

Therefore you can't expect that an external expert will be able to implement a new, selected project management methodology off the cuff. Firstly, in order to select a methodology, you need to analyze which of them will best suit your organization, think about why we want to introduce these changes in the first place. Secondly, an external expert will need time to understand the specific situation of your company. They will need your people to explain it as they know most about it. Very often the client requires that a project is executed in a particular way, for example, using the agile approach. It's a whim. Then it turns out that neither the supplier nor the client is able to work in this methodology as it would require some fundamental changes in both organizations. Transformations like this take years, and our project is planned for, say, a year. In the end, both organizations are struggling to solve problems resulting from an inappropriate methodology instead of focusing on the project and business goals.

A project is not a testing ground for people who want to learn a new project management method. The client has some concrete business goals to realize and that should be the main focus. Project management methodology should be selected in order to ensure the fastest progress, while taking into account the conditions in which the project is executed.

Summary

- None of the popular project management methodologies or methods is a full remedy for all problems and it also does not guarantee a success.
- The choice of method should not be dictated by its popularity or reviews. The decision has to be made on the basis of a thorough, multi-faceted analysis of the project's environment.
- You shouldn't be ashamed of creating your own solutions. Methodologies and method are just a set of good advice that you can learn from. You should choose what can potentially be effective and eliminate the elements which will not work and might even hurt performance.

- During the project, you can't follow any methodology blindly. It's necessary to closely monitor the situation, get rid of practices that are ineffective, implement improvements. Our goal is to carry out the project for the client and not to thoughtlessly follow guidelines from some coursebook.

CHAPTER

14

The Management Strategy Depends on the Stage of the Project

Insanity is doing the same thing over and over again and expecting a different result.

—Albert Einstein

M. Dąbrowski, *Managing IT Projects*,
https://doi.org/10.1007/978-1-4842-9243-3_14

Depending on the stage of the project, different aspects are emphasized. During the sales and initiation phase, the competencies and experience of several key people are stressed, and relations and trust are built. During the analysis stage, the work is mainly done "on paper" and in a virtual way. It's still important to demonstrate knowledge and experience; a handful of people are involved. Then, when the development starts, we move to a phase when the client cares about concrete outcomes, we focus on numbers, performance, velocity. Depending on the selected methodology, the client is more or less involved in giving feedback and creating some final solution. Relations with the key analysts are less important than the outcomes of the development work. The closer we get to the stage of UAT, the more we focus on the discussions related to what is in the scope and what isn't. Individuals don't matter (that much) anymore, the client can see the software that was delivered and directs his entire attention to "polishing," which means ensuring that nothing is missing. Thus, it's a period when both businesses look at the completeness of the delivered solution and how possible fixes will affect the final release date. Then the solution is deployed into production and there is time for stabilizing the system, the businesses start talking about SLA, the number of bugs in time, bug category, fixing times, etc. Nobody talks about the pace of development, the company's competences, the number of people on the project. The client is only interested in quick fixing of the bugs, and even better that there are no bugs whatsoever or that their number does not exceed a certain limit. So, depending on the stage of the project, both the client is interested in monitoring other aspects of the project, and we also need to emphasize the right things and avoid talking about some other (for the sake of the project and of the client), while some are just ignored altogether.

Sales Process and Project Initiation

At this stage, it's a key thing to demonstrate the people's competences and their availability. There are no problems typically encountered in project work yet because they surface over time. Thus, we focus on showing the client our best "players." Their main task is to skillfully handle the client's questions, create a vision of the system, of the product, and business processes, and – what is even more important – to build relations with the key and most decisive people on the client's side thanks to which the supplier can first get the contract and then shape it in a beneficial way, modify it, and make some crucial assumptions in terms of further development.

The main problem that the supplier has to face is availability of the key people mentioned previously, as these are usually assigned to other ongoing projects. Making them attend all meetings with the client during the sales process usually ends up in internal conflicts and escalations from the current clients. Then it's necessary to assign these people to the newly acquired contract, which leads to even more chaos and sour feelings on both sides. We need to be prepared for that.

It's important to remember that it's at this stage that all the "incurable diseases" of the project have their source. That's why it's a mistake to treat the sales process and the initiation phase as a happy time when nothing can go wrong yet. Admittedly, that's most often the case. It often happens that suppliers, excited about getting a new client, turn a blind eye on the things that are decided about in the first weeks of the project. Then it turns out that some analyst agreed to the client's requirements in some domain and it increases the scope by a hundred man-days. It happens that the supplier lets some much extra work slip into the scope that the project becomes unprofitable.

Analysis

Regardless of the methodology that we decide to adopt, there's always analytical work to be done. Here it's also people, their competencies, and experience that are most important for the client. In this period, the client looks at the supplier through the lens of the analysts assigned to the project. If they are good, experienced, communicative, persuasive, and managed to gain the client's respect, he will be pleased despite some minor delays. At this stage, the most important thing is whether the supplier understands the client's expectations, whether the communication between the members of the project team is seamless, etc. Even if the supplier has to start production of the system at the analysis stage because there's some concern about delivering on time, the client is still unaware of the issue. He focuses on the analysis, documents, concepts, and agreements.

During the analytical stage, suppliers often "distribute the sold delay." If – regardless of the developers' protests – the supplier sold a project with an unrealistic schedule, the analysis phase is an opportunity to play on time. It is at this stage that dependencies between the client's tasks and the supplier's tasks become clearly visible. Most often at the beginning of the project, the client has some problems with mobilizing people and getting the necessary information. Clients pay the supplier for doing the work so they're hardly ever prepared to face so many demands put on them. Thus, the supplier can then expose the resulting deficiencies, delays in providing the data, documentation, the client is blamed for postponing decisions or for not being able to reach an agreement regarding a new way of working. Obviously, as a supplier, you need to know the limits of that blame game.

On the supplier's side, on the other hand, the main problem is to build a strong team and then maintain it. If the project is challenging, objectively complex, there is a new product to be developed, we need the best people. If some competencies are lacking in the organization we definitely need to acquire them from the market. The attempts at analysis made by people who are not there yet, are not experienced enough, or lack domain knowledge is usually asking for trouble – either the clients lose their patience because the level of discussion is unacceptably low or it is cleverly used to add new requirements, or it's the supplier's representatives who can't persuade the client to their interpretation of the scope, which consequently leads to a substantial increase in costs and delays.

Development

Regardless of the management method, at some point the main weight of running the project will be moved toward the development effort. With time, the client will also focus more on the outcomes of the development work. They will be interested in seeing some tangible outcomes, prototypes, and demo versions; more review sessions will be held to get feedback on the "live" system. In this period, most emphasis is put on monitoring progress based on hard data, regardless of the form of its presentation. Clients tend to assume that they had already communicated their expectations so they pay less attention to competencies or allocating specific people. They still insist on having the best team, but it's not as important for them as it was in the beginning of the project when the requirements, scope, and goals were discussed. At this stage, they expect outcomes. For them, these are two simple things. On the one hand, they want to make sure that the development is going according to plan, which can be measured by various tools specific for a given management method – the status can be presented in relation to the schedule, man-days used, or delivered story points, it can be shown in tables or on diagrams. On the other hand, clients may not trust the data in paper reports, so demo sessions need to be held at regular intervals.

From the supplier's point of view, the situation begins to polarize at this stage of the project. To put it simply, internal progress is what should be reported to the client.

The main problem is to reach the anticipated pace of delivery. In order to achieve that, the right environment must be set and the right team with the right competencies and capacity has to be built (and then maintained).

In theory, the management strategy seems simple and clear: we need to create an effective team and then let it work, while removing all the roadblocks on its way. No doubt there will be some objective difficulties. There will be other projects which might potentially need the engagement of some of your team members. There will be technical problems which hadn't been anticipated. Some work will be blocked either by the client or by some dependencies between particular people or teams, etc. However, if the problems will be happening for objective reasons and will be the result of neither bravado nor neglect, if you're able to show the client that you proactively and pragmatically work to find a solution, you're doing everything you can, the project should be safe.

User Acceptance Tests

At this stage, what counts most is the quality and the determination to reach the planned level of quality. It's obvious that the tested product or system should be as faultless as possible. However, normally when we work under time pressure, we're striving to meet ambitious deadlines, the project is complex, dependent on multiple actors, etc., eventually we end up with some quality issues. Then the main concern for the client is whether the process of bug fixing is deterministic, whether it can be planned and whether the supplier has it under control or not. If the system delivered to the client has quality issues, but we are able to improve its quality within a set time frame and prove it with some objective measures, clients usually accept this kind of situation. It's much worse when, over time, the number of bugs does not decrease and we can't come up with a reasonable explanation for that. It can very often be traced back to hurried, chaotic, and parallel development of many interdependent components and neglecting quality checks on the way. The bigger the system, the more complications like these.

The matters relating to competencies, the sufficient number of people, particular key persons, become less important for the client. He has verified if the system meets the preliminary expectations and now he's merely waiting for the right level of quality to be ensured. On the other hand, it generates numerous unanticipated problems for the supplier. It's often the case that the estimated cost of quality assurance is much lower, people are planned for other projects or some of them are moved to work on the next stage of the same project while user acceptance tests are being performed. And then there is disappointment, the team still needs to be engaged in bug-fixing process, other projects are understaffed and in our project the start of the next phase is postponed until the completion of UATs which seem to never end. That's why the matter of estimating the length of the approval process and the costs of quality assurance should be treated very seriously. Yet another problem is if the client wants to listen or understand that the project must take longer than it was originally planned and whether he accepts that.

Stabilizing the System After Going to Production

Right after going to production comes one of the most challenging stages of the project. Most people think that UATs are the worst, but unfortunately, it's not the case. Tests can be carried out in a relatively controlled environment. You can plan, delay, persuade the client, buy the time, etc. But when the system goes live and real end users start working with it, the situation changes dramatically. The client's team, which up to now held the project under control, lose it on the spot. The so-called business comes into play and these people want to achieve their goals no matter what. The time of understanding, mercy, and amicable dialogue is over. The most important thing is that no critical bugs can happen in production, and ideally, there should be no bugs at all. The supplier must set up the right team responsible for maintenance and mobilize the right number of people who will have to work in very tough conditions, under enormous stress, who will have to drop any activity and fix any bugs which block the client's business. Even one serious bug can alert the entire supplier's organization! That's why it's so important to set the right escalation path, a plan for effective communication and teams of people who will always be on duty, always ready to take action in an emergency situation.

Maintenance

After stabilizing the system in the production environment, finally there comes a time when the supplier's work can be anticipated and planned. Very often, it's the most lucrative stage of the contract. Of course, the expected quality has to be ensured, ideally, if it's perfect or at least meeting the requirements set in the SLA. It's enough to set up the right competent, efficient, and preferably dedicated team if the conditions allow that. It must be remembered, though, that your approach to the client might determine the further development of the business. If you see the client only as a "problem," the party who is never satisfied and always reporting bugs, the cooperation will gradually deteriorate and will probably come to an end when the contract expires. Whereas, when you treat the client proactively, you talk to him on a regular basis, exchange experience, help in the adoption and use of the implemented system, demonstrate new product capabilities, the cooperation will thrive, the client will feel looked after and he is likely to order new products from you and start new projects.

Summary

- The way of managing a project evolves along with the project.
- Initially, the human factor is the most important: people's experience, their domain knowledge, the ability to build relations. It translates into easier and more accurate requirement analysis, creating a bond with the client, building trust. At the later stages, hard measures start to count more, for example, performance, reports, and numbers. At the very end, statistics, speed, and predictability in solving problems are what become most important.

- In line with the changing stages, we must adopt a particular management style, focus more on some specific areas, and in particular pay attention to what should be exposed to leave the client with a sense of control, influence, and being understood. In the first month, it's no point boasting about how fast the system is being developed if the client isn't exactly sure yet what he really needs in order to achieve the business goals.
- Regardless of the stage of the project, you need to be proactive, support the client, and be ready to help him.

CHAPTER

15

People's Availability at Particular Stages of the Project

Everything put together falls apart sooner or later.

—Murphy's law (Simon's Law of Destiny)

One of the most neglected matters in project management is the stability of the project team. It is assumed that at the beginning of the project, there is a stage of building a team after which you smoothly move on to discussing management matters. A lot has been written about scope management, planning, monitoring and inspecting work, risk, quality and time management, etc. There are numerous project management methods and methodologies. They're all built on a premise that not only there is a team but also that it consists of skilled and experienced professionals with the necessary domain knowledge.

M. Dąbrowski, *Managing IT Projects*,
https://doi.org/10.1007/978-1-4842-9243-3_15

In practice, however, it's an issue for a whole separate publication. Building a project team or at least its core – the key individuals – is a remarkable feat for IT companies. Nobody keeps such people idling. Each of them is already involved in some kind of undertaking. The best people are also the busiest ones and they are a part of companies' most ambitious projects. It's relatively easy to make a mediocre team consisting of the newly hired, inexperienced individuals who lack the knowledge about the company's inner workings, about the client, technology, or products, whereas the best people are extremely hard to get. It's related to both internal escalations as well as resistance and threats from the existing clients.

An IT company is an ecosystem in which new contracts constantly emerge and almost everybody on board is engaged in ongoing projects. Building a team for a new client is far from easy. What's more, due to the fact that the company keeps selling, the project team will inevitably change. At the beginning of a new project, some people will have to be "stolen" from other departments or projects. This process is hardly ever frictionless. After some time, some of our best people will also be gradually moved somewhere else, as there will be new sales processes going on or some projects will call for emergency solutions and apparently only someone from our project will be able to put out that fire.

To make a long story short, it's necessary to assume that our team will keep changing dynamically throughout the project. If we fail to understand the mechanisms at the company, we will be surprised every time, and we will also be the ones to blame. Due to the scarcity of resources, it's possible to distinguish the phases in which we can expect some management decisions to take place. If we understand it, we can prepare for them in advance.

Sales Process

In reality, the project starts as early as the sales process, which is often a year or two before the signing of the contract. At some point, we get to very serious business talks. The client's representatives want to understand the features, it's necessary to convince them that our offer is better than any other, and also to explain why they should pick ours. In order to draft a good and convincing technical proposal it's crucial to understand the system environment on the client's side. What are the bottlenecks currently? What isn't working? Which elements should be replaced and in what order? What will the gain be at each stage? Besides, in order to estimate the price of implementation, we need to estimate the cost of integration, which entails also checking the interface map for various tools in the client's ecosystem.

It's difficult to carry out all these detailed analyses if you involve only sales people and pre-sales consultants in this process. For this reason, most businesses create some kind of a project team consisting of the best consultants, the most experienced developers, and implementation specialists at this stage. Very often, a preliminary decision about the future project manager is also made.

The team's task is to understand the problem at hand and to estimate the implementation costs but also to show the supplier in a positive light. It's said that it's more difficult to get a new contract than to carry out a project. There is a lot of truth in this statement. That's why the initial team is the company's intellectual crème de la crème. In fact, it will often be the best team that the project manager will ever have a chance to work with (on this particular project). It will be a sort of a "dream team." These are the only people who can convince the client to choose this particular supplier. Pre-sales consultants are wonderful at oozing charm and presenting. Developers and implementation experts are able to deal with any technical question, and what's more important, they will ask questions which will make the client believe that he is choosing professionals who know their trade.

Initiation and Getting the Project Up to Speed

After the contract has been signed, the entire knowledge about the coming project is in the minds of the project team created at the sales stage. The contract documentation only contains a subset of information without too many details, without the key insights, without some crucial internal assumptions on the execution, without the information about the key risks and pitfalls. After all, you can't put it in the contract and make it known to the client.

Thus, it's a very reasonable thing to build the project team around the core of the one that was involved in the sales process. Even if it means that the pre-sales consultants are held back from other projects for several months, that's the way to do it. Whether a project can succeed is largely dependent on the flow of information and access to it. First of all, we must ensure that the knowledge transfer from the sales stage to the implementation stage is smooth. Another problematic issue is also that the best developers and implementation experts who had been temporarily involved in the sales process. As a matter of fact, they also had their ongoing work to do in other projects. As it is more difficult to manage a new implementation than it is to handle an already stabilized project, these people should also be transferred to this new undertaking. At least for a few months, until the knowledge has been shared, the project has gained traction, the right relations and ways of working with the new client have been established.

So, the first months of the new project are a period when we have the best team that we will ever have on this project. After that, it will only be getting worse. It's only a matter of time before there will be a more promising client, a more lucrative business on the horizon and the consultants that were temporarily involved in the project will be taken away without a word of warning. It will also happen to the developers and implementation experts who will have to return to their duties at some point. That's why, from the very start of the project, you have to meticulously plan who will be on the project management team, and especially leaders of specific areas, key analysts with the emphasis on the people directly involved in the client cooperation. It's key to pay attention to the knowledge transfer and be prepared for this change. Resources will be moved and nobody will be concerned about how hard it is, how it affects the project and also the existing client will likely escalate. To make matters worse, we will be the ones to blame for all this mess. The only reasonable solution is to plan ahead and prepare for this situation.

When the Project Still Has High Priority

When the project starts and gains traction, which usually happens after a few months, the team created at the sales stage is gradually dismantled. As new clients need attention, we need to "return" the pre-sales consultants. The experienced developers and implementation experts, borrowed for the period of acquiring the contract and starting the project, go back to their domains, departments, and problems.

There comes a time to work with the final team. This stage lasts until our project has an internal priority, in other words, until the relations with the client are important and may bring potential new business, and most importantly, until the company is making a profit, the invoiced amounts are high, higher than in other projects.

It's hard to state precisely how long this situation will last. Usually it's a year, maybe two, sometimes a bit longer.

At this stage, we might no longer have access to the best people in the organization (those which are needed when new contracts are acquired), but we usually have a good, stable team. Until invoices are issued (or at least are likely to be issued), we shouldn't worry about lacking resources. Thanks to this, we can consistently do the work, and the organization shouldn't stand in our way. It's obviously a simplification.

Understandably, there will be instances when it will be necessary to let some people go so that they can enchant new clients, urgently "put out a fire" in another project, etc. However, then we still have a chance to fight for keeping the team, as we have the project's profitability as an argument to support it. If others keep stealing our people we can escalate, stating that we won't be able to invoice a particular amount on time which will negatively affect the annual bottom-line. We may also use the threat of client's escalation as an argument. These strategies will only work until the amount to be invoiced is still significant and the client's rating within our organization remains high enough.

When the Project "Ages"

After that, it gets only worse. After some time, the project loses its status in the organization. The client doesn't seem so attractive anymore, the management is no longer excited about the new catch. Especially when the highest amounts had already been invoiced, the profits on the project are mediocre, poor, or just unimpressive. The more skilled individuals are systematically moved to other, more lucrative projects or to those dealing with more serious issues. All this happens in the vein: *"it's more difficult to get a new client than to work on the 'cozy' and already stable project."* Managing a new, starting project is also objectively more complex. It's necessary to build a relationship with the client, show your competencies, prove your experience, convince the right decision-making people, and gain trust. We can achieve all this only by relying on a team consisting of the best people. That's why when the project has aged, is stable, the relation with the client is mature, the client is understanding, it is unfortunately (however rightly) assumed that our project is the one from which people can be taken away and that's what it will be like until the end of the project.

The faster the company is growing, the more aggressive its sales strategies are, the more often we will find ourselves in a situation when the best people are taken away. We could say that the longer the project has lasted, the more mediocre people we have on the team!

Besides that, the fate also reaches the project manager himself. He will also have a hard time to reach the end of the project that he initiated. At some point, the organization will need to fulfill the management positions in a new, more challenging implementation project. Who will they reach out for? For newly hired individuals or maybe for more experienced project managers who had been hardened in this particular organization, they know its structure, individual directors, team leaders, and specialists who know how to work in this very organization, understand the project management methodology, can talk to clients, know the products, etc?

At this stage, our project is a place to get resources for other undertakings and some sort of a place for honing talent. Since the situation is stable on all fronts, the relation with the client is perfect, there is mutual understanding, we can reliably plan and deliver, this will be the ideal environment to learn for the new hires, the place where they will start off and acquire all the necessary skills. Obviously, before they get "stolen" for other promising or challenging undertakings.

It's perfectly clear for clients that the prerequisite for succeeding in projects is having a strong and competent team. What's more, it's essential that the team remains stable throughout the project. As a result, it is ensured that the arduously acquired knowledge is kept, the relations built between individual team members are maintained for as long as possible, the team works in a predictable environment and remains effective. Any changes on the team cause disruption and are generally harmful. That's why some clients insist to have in the contract some mechanisms aimed at protecting their interests, for example, they want the key individuals on the supplier's side to be specified, they might also require procedures for changes on the team, and the client has to give consent for each change.

Summary

- You need to realize that a project team constantly changes during the project. If you understand the processes in the organization, it's possible to prepare for these changes.
- The first, initial team is created at the sales stage. It usually consists of pre-sales consultants and selected, best developers and implementation experts. This is usually the best team that the project manager will have a chance to work with.
- During the initial stage and in the first few months of the project, it is reasonable to keep the initial team intact in order to later distribute knowledge in the target team, bring the project up to speed and show the client the competencies, and lay the foundations for a relation.
- At the later stages, the team created for the sales process is gradually dismantled and individuals are replaced with the members of the final team.
- When the project still has a high priority in the organization, the client still has potential and the invoiced amounts are high, we have the necessary resources and most often the key people are not taken away.
- When the project "ages," it has lost its high priority and is stable, we get into a phase in which the project is treated as an incubator and a training center for people who will be later involved in new projects. We need to deal with constant replacements of the best people with the newly hired ones who need to be taught everything.
- The longer the project lasts, the weaker the team the project manager has to work with.

CHAPTER

16

Monitoring the Internal Status of a Project

Most of what we call management consists of making it difficult for people to get their work done.

—Peter Drucker

It's much too often that we yield to the temptation of avoiding the internal status monitoring for projects. We have the reports for the clients and meetings with them, so why would we add more work and annoy the people who – in their own words – "want to focus on doing their work." Then most of the effort goes into making impressive slides for the client. The graphic part and the aesthetic aspects are placed higher than the informative value of the

M. Dąbrowski, *Managing IT Projects*,
https://doi.org/10.1007/978-1-4842-9243-3_16

presentations. We end up with pretty slides, tables, diagrams, colorful lights that the audience finds confusing and misleading. The relevant information is not presented in a straightforward manner and the reader must put in a lot of effort into figuring out what it's all about. Project managers try really hard to hide uncomfortable facts, skillfully avoid explicitly calling problems and risks – all that is caused by fear and embarrassment in front of the client.

However, what's worst is that preparations for the status meeting with a client, especially with the steering committee, takes so much time that people get sick of reporting for the next days or weeks. That's exactly the reason why internal status meetings are skipped. The report for the client has just been issued after all. The problem is, though, that it's a serious mistake. These are completely different reporting dimensions. If a restaurant guest asked a waiter what the chefs exactly do to make a specific dish, would he be satisfied with the answer? I suppose he would lose his appetite instantly. Similarly, the report for the client is as detailed as the waiter's answer to the guest's inquiry. If anyone wants to manage a project based on the information presented to the client, they likely won't get very far. Internal status meetings are key for success. They are the space to talk about actual progress, risks, problems, and their solutions.

Two Parallel Reporting Systems

So, we have two completely different reporting processes: for the client and the internal one. Each of them has different goals, puts emphasis on different aspects and is clear for another audience.

The client is interested in a general status of the project, especially in relation to the original plan, and this includes any deviation from the schedule, delays, scope creep, and how it affects the budget. The supplier should mostly focus on the problems that are directly related to the client (cooperation, analysis, expectations, infrastructure, deliverables from and to the client) so that the work goes smoothly and is not blocked by the client. It's also important to care about the company's good reputation, the opinion about the project, maintaining the positive PR. It's necessary to give an impression that even if there are problems with the project, they are under control, we know what we're doing, we're consciously carrying out the plan, nothing is left to chance. We should avoid talking about the internal problems, chaos if it's the case.

Even if there are various difficulties at some point, showing them to the client will not only fail to help in anything, but it will also create a climate of uncertainty, will breed doubt and will make the client consider other options as he will start to doubt whether the right supplier had been selected for the job. Creating a positive image of the project, its state, and our organization is very often neglected. If most projects are delayed anyway, which of them stand a chance to last till the end? Those in which the clients are demotivated, discouraged, and annoyed or those in which they have a sense of security, control, influence on what's happening, and get the supplier's support?

Internal reporting has completely different aims. It's focused on progress, monitoring the performance on the supplier's side. It concentrates on the organization of internal work, its efficiency, eliminating waste, it's about risks, problems, and especially on the ways of solving them.

While reports for the clients selectively present problems related to the cooperation with them, internal reports are mostly focused on problems, removing impediments and roadblocks, any lacking resources, conflicts, ineffective processes, tools, etc.

When we talk about the project status with the client, we try to care about PR, we often avoid some negative aspects while emphasizing the positive ones. What matters in internal reports is only the brutal reality: truth, truth, and nothing but the truth, and also data, data, and nothing but hard data. If we know the facts without sugar coating, without avoiding tough issues, we are more likely to deliver on time or as soon as possible. It's a common problem not to be able to get a real progress report from your own people. Unfortunately, everybody has their pride, is afraid of something, wants to keep face, feels embarrassed. Certain individuals may hide difficult issues, report only positives, which is obviously counterproductive in relation to themselves, to the project, and the business. That's why it's so important to create an environment of openness in which everybody can talk about any issue, regardless of how difficult or uncomfortable it is, and they will never be punished for that or ridiculed. People should be encouraged to talk about problems and offer solutions. Internally, it's crucial to monitor performance, throughput, and velocity of particular teams. In an environment in which multiple projects are executed at the same time, it's key to constantly monitor the capacity in order to be able to assess if we can deal with a certain scope in a given time.

It's absolutely necessary to measure at regular intervals the pace of delivery of certain features, regardless of the applied methodology, measurement units, terminology. What counts is the regularity and using hard data!

You can't just rely on wishful thinking. If the data shows that the pace of work is too slow, there must be some problems at the core. It might mean that the team is too small, ineffective, there are some roadblocks, people don't understand the requirements, the environment is unstable, there are dependencies between teams, etc. It's crucial that the measurements are based on numbers and not on subjective descriptions or opinions of particular individuals.

It's the Information That Counts, Not Its Format

A useful internal report has nothing to do with the selected project management methodology. It can't determine what information will be presented and in what form. A good and reliable report is not some colorful slides or sophisticated formats. Too many people waste time "polishing" the elaborate templates which nobody either understands or uses. It leads to frustration in people who must work with these "useful tools." What counts, what really matters is information, regardless of the form in which it's presented! What matters is that the data is complete, also that it can be compared to previous periods.

As long as we're getting useful information we may let the team issue reports in any form they find convenient and fast.

On the other hand, report templates should be used, but absolutely not because of their aesthetics or format. They need to be treated as some sort of lessons learned and a checklist. We're talking here about templates that have been worked on for years and not about those which a project manager has made based on a popular book or a training that he's just attended. If there are useful tools for reporting in our organization, we should approach this issue with humility and think about why it's like that. While we're preparing the status data, templates force us to think if we have taken into consideration each and every important aspect of the project.

Good Practices That We Constantly Forget

While the format doesn't matter, there are some good practices, "golden rules" that should be always and relentlessly followed both for preparing status data and during a meeting in which the status is discussed with the team. They are independent of any project management method or methodology – they must be always followed! They are as follows:

1. Each task has its owner – there can't be any "ownerless" tasks. There must be a specific person responsible for every task. This individual must know about it and take ownership of the task, otherwise there won't be any progress. It's unacceptable that a question about the status of a specific task remains unanswered. It's a symptom of pathology.

2. Each part of the scope has its owner – whether it's an element of the system, functional or competency area (like tests or data migration). If there's any doubt, problem, or risk in a given part of the scope, it must be clear and evident who should be working on the solution. What's more, this person should act proactively, show initiative, and not wait until they are called on.

3. Each risk or problem has its owner – otherwise we'll experience the shaming and blaming game. If a problem or risk comes up in a project, it must be ensured that there is someone accountable for dealing with it.

4. Each task, risk, or problem has a due date – we can't allow the team to talk about the problem without a time frame. In a week, the problem is brought up again and then again, and so weeks and months pass. It's simple and clear: either there is a problem and it needs to be solved by a certain date, or it will lead to serious consequences; or maybe there's no problem if it can be postponed without any serious consequences. In other words, if the problem is serious, it must be eliminated by a certain date!

5. Deadlines must always be met – very often in reports, deadlines for tasks or solving problems are postponed: in every report or status meeting. These are not some irrelevant numbers! The project schedule consists of individual, small tasks. If we carelessly allow for the deadlines to change, we'll eventually end up with delays that are visible for the client. For individuals, developers, or teams, the issue might seem unimportant, but in reality, it's extremely serious. This is exactly how project delays are generated. That's why it's necessary to build a culture of meeting deadlines and committing to own estimates.

6. There is always one person accountable for a particular deliverable – whatever we're working on: a part of the system, project schedule, analysis with the client, plan, or carrying out user acceptance tests, migration, integration, etc., it must be absolutely clear who is accountable for delivering on time. Any reports in this area must be written by these people. In meetings, we should talk to these individuals. They should be the ones who have a sense of urgency. They should care about the deadlines, about finding solutions to problems, about talking to the client. It's unacceptable that the project manager has to

tell everyone what they should do – and unfortunately, this is most often the case, the project manager is the only one who is pressed for delivering, and the rest of the team is somehow relieved from this responsibility and sleeps well. Pressure and responsibility must be distributed among all team members. Everybody should be accountable for their part of the project – proactively trying to achieve the goal without being controlled or pushed. Only then can a project be successful.

7. Accountability is assigned to one person – a basic mistake is to assign accountability to a group of people. It inevitably leads to chaos in which – when there are difficulties – each of these people hides behind others, blames them, and claims that their responsibilities were unclear from the start. That's why only one specific person should be accountable. Of course, everybody on the team must know who does what.

8. Who, what, why, when – whether we're preparing a report, facilitating a meeting, we should always use factual data. We need to communicate clearly and expect the same from other team members. Each piece of information must be described by "four dimensions" (who, what, why, when). It doesn't matter if it's a task, problem, risk, the need to check something. Each activity in the project, any task that we can think of has an owner who should understand what they need to do, why they need to do it, and by what date. Contrary to what you might think, it's not that obvious, at least not to everyone.

9. Avoiding problems doesn't make sense – for some unclear reasons, a lot of people prefer to hide problems, say things that are only partly true, avoid them without understanding that an unsolved problem generates other problems, which gradually leads to complete chaos. The simplest, the most effective, and the smartest thing to do is to face the problem – fast, without delay, without wishing for a brighter tomorrow. The main "pain" and effort is usually connected with the decision to take the bull by the horns. Then it usually turns out that the solution was simple, quick, and didn't require so much effort. The art of confronting difficult situations is a habit that can (and should) be developed both in yourself and in the project team.

10. Each time it's expected that outcomes (progress) will be presented in relation to the previous meeting (report) – it's extremely common that the task status Is a narrative that remains unchanged for several weeks. Some people can actually generate reports by "copy, paste" method because it's convenient. If people report like this, there is an illusion of control and progress, while in reality, there are delays, blockers, and problems. That's why it's necessary to clearly show what has been done since the last meeting or report, what specifically has changed, and where there is stagnation.

11. If you have a problem, how are you going to solve it? What have you done so far? You should expect this kind of attitude from everyone, even from yourself. You can't allow for situations in which somebody reports a problem and just waits. Who should solve it? Project manager? Are they supposed to deal with everything? Why does the team even exist then? Or should it be the "others"? This attitude leads to a disaster. Each member of the team must fight. Even if other people are supposed to help, he is the one responsible for motivating them and ensuring that it happens. Everyone has to be proactive. After some time, all team members get used to that complaining is not enough, that it's necessary to act, that everyone can and should do something. It changes the work dynamics completely.

12. Moving problems back and forth is a waste of time – this sort of behavior needs to be eliminated at all costs. It's common that the reporting person claims that they are blocked due to waiting for "others" to finish. However, they haven't talked to those "others," haven't told them how important the tasks are. So, how are those "others" supposed to know that they are blocking someone's progress? They have their own tasks. Then it usually turns out that it was enough to spend five minutes and talk to someone and the problem would have been solved. But it's obviously easier to blame "others." That's why it's essential to model how to act proactively and relentlessly eliminate bad habits in every meeting. Usually, after some time, the team becomes proactive and starts eliminating this kind of dysfunction.

Who Needs Status Information and Reports Anyway

At the project meetings, there are often questions in the vein of: "What form do you want it in?" "When should the report be ready? Is next Wednesday okay?" "Is this report detailed enough for you?" On the other hand, team leaders or scrum masters asked about progress can't answer straight away and say: "I need to check that, I'll prepare it for tomorrow." It is a symptom of a widespread misunderstanding as to what status meetings or reports are for in the first place.

Reports are not written exclusively for the project manager, department head, or for the client! First of all, it's a tool for all people responsible for delivery monitoring at each management level of the project.

It doesn't matter whether it's a leader of a team consisting of three developers, a scrum master, someone responsible for migration work or test coordination: reports and status meetings are especially for them! They are the ones working on the project and that need to know what they should do, when, how far they've come, and what problems they are facing. But reality is often quite different. A department or business unit head talks to the project manager about progress, they talk about it to the client, but if you went to the team leader instead, to the place where the actual work is done and asked him what the plan is, what the expected outcome is and when it is due, what the milestones are, it might turn out that he has no idea! In this case, it should be seriously considered what the reports for the project manager are based on.

Project management can't be a situation in which the project manager runs from one person to the next and infers what the progress is on the basis of some incomplete information. Reporting is a process that must go through all management levels. Each person responsible for coordinating work must prepare a report for themselves to understand where it's going and what the current status is! Reporting is not the project manager's job, it's every team member's job. Thus, when a team leader or scrum master asks the project manager: "When do you need this information about progress and in what form?" they should answer: "I'm asking you when you need the information about your team's progress. When would you like to know what the performance is and what problems you've had?" "What form do you think will be suitable to be able to monitor the work?"

The team needs to be constantly reminded that the status information is for them, not for the project manager. The project manager only ensures that the people have the right tools to monitor the status and use them actively!

In an IT project that 150 people were working on, the supplier was dealing with deadlines being notoriously postponed. In theory, every week the project manager prepared a detailed report about the progress, both for the client and internally. As it is with this number of people, the management structure was multi-level, it was decided to check the leaders' knowledge about the status as it was they who were accountable for delivering particular components of the system. They were asked for reports with some basic information about scope which included

–Five major features that were important for the client

–Main milestones that the leaders were to achieve with the teams

–Five major problems that the teams were facing

–Five most important risks that the teams were dealing with

To the project manager's surprise, the leaders weren't able to provide this information in the meeting! They needed three weekly meetings, so altogether a whole month before they could report the status of their own teams.

The preceding example shows that in most projects, we follow an illusion that if the project manager prepares the reports for us or for the client, the project is being managed. Absolutely not! It might be that we're reading some colorful reports, and below the surface, there is complete chaos, people don't know what and when they should be doing. And obviously we can (and should!) ask whose fault this is. Thus, reporting should be seen as a process and people should be made aware that reports are not for the "management" but for ensuring that the teams and the work are being deliberately managed. Status information is for them! It's their tool! Reports are made mostly for the teams themselves!

Summary

- It's impossible to manage a project based on the information prepared for the client. Internal reporting is key to delivering on time.
- The aim of reporting is to give information, not to engineer slides. The form doesn't matter. The data matters.

- Regardless of the project management methodology, you should bear in mind that there are some basic rules for monitoring progress such as appointing the accountable person, meeting deadlines, demonstrating changes in relation to the previous report (meeting), dealing with problems, etc.
- Reporting is a process that should go through all project management levels. Status information is useful for everyone, not just the project manager.

CHAPTER

17

Low-Level Estimates – A Source of Everlasting Delays and a Foundation of Planning

Compared to the ability to organize your work on a daily basis, everything else in life seems just childish.

—Johann Wolfgang Goethe

M. Dąbrowski, *Managing IT Projects*,
https://doi.org/10.1007/978-1-4842-9243-3_17

When delays finally emerge in a project (and it is just a question of time), the process of hunting the guilty party begins automatically and in some sense subconsciously. As it turns out, in this matter, most organizations display here some sort of a Pavlov's reflex. If you ask anyone who is to blame the "obvious" answer is: project manager. It's convenient for everyone. The management has a scapegoat, department manager's hands are clean, project team members are relieved. Thus, the following faults are blamed on the wretched project manager:

- Can't prepare a decent plan
- Can't monitor progress
- Is incompetent at managing people, does not execute, does not push them, does not convey the sense of urgency
- Does not see the dependencies between individual teams
- Is generally disorganized and unwise

And the problem's solved. But is it really? It can indeed happen that the project manager is incompetent. But is it really possible that it's the main cause in most cases? If it was so, it would be enough to invest in a training program for project managers and all software companies would be delivering projects on time.

It's worth wondering, though, what it means exactly that a person cannot make a proper plan or monitor progress. How are such plans created? Where do the data, specific numbers, dependencies, and deadlines come from? Are they provided by the project manager? Or do they get them from the team and some domain experts? What is the management process in this particular company? Do all one hundred people involved in the project report directly to the project manager? If not, how can they influence the work of these people and meet the deadlines? Such questions are virtually endless.

In practice, the project manager is the most competent person in the area of planning, monitoring, and controlling the work, in other words, in project management, so it's only logical that delays cannot be the result of their mistakes (even though some mistakes may occur). In reality, the planning process is faulty at the lowest organizational level, at the stage of estimating performed by individual team members (low-level estimates). Each team member estimates their own workload and assumes that they will finish the task by a certain date. After that, it turns out that the work took, say, twice as long.

It's common to hear excuses in the vein of: "the problem turned out to be more complex" or "someone misled me and I needed more time" or "the management's expectations were unrealistic." In general, such information is not shared wider. What for? It's only a small part of the project, one of many tasks of this particular specialist. However, a project is a whole comprising these smaller units of work. Thus, eventually the sum of these supposedly minor delays results in major delays within the whole project. The worst part being that nobody understands these delays! There's no clear answer as to what exactly happened. There's no one specific cause that could be presented to the client or the management.

In this chapter, we are going to skip all the obvious situations such as glaring management mistakes or executives' incompetence. Such things happen, but they are so evident that describing them is pointless – there are numerous books and articles about it.

The Scale of the Problem

Let's imagine two hundred programmers, and each of whom claims that completing their task took twice as long in relation to the original estimate. Usually it translates into months in delays. How can we explain to the client what happened? Maybe the organization has some management issues. Or maybe the problem lies in individuals.

Interestingly, the most prominent excuse repeated in such a situation is along the lines of: "That's how you sold the project, we can't uphold this unrealistic schedule." Then the team is usually asked to come up with a more "realistic" plan – such that the people believe in and which will be executed. Of course, when this ideal schedule is crafted, it's presented to the client for approval. This process is obviously incredibly hard, and it consumes a lot of time and effort. There are often negotiations, fights, threats, and contract breaches. The management apologizes, gives promises, and claims this situation will never happen again. The people made the new plan themselves, worked on realistic estimates and assure everyone that (this time) the work will surely be delivered on time. Later though, as usual, people don't meet their own deadlines claiming (once again) that "they had been forced to agree to the unrealistic estimates." And so it goes again.

There are also some other delay generators which directly affect the delivery time of individual tasks in such a way that it's difficult to explain the reasons. These are among others:

- The scope of the project described in a very general way, without details, which gives the client the possibility to interpret it in any way he pleases throughout the project
- A different view of the scope for the client and the supplier

- Analysts lacking sufficient domain knowledge and the characteristics of the product to be delivered, the organization they work for, and for which they deliver, without enough practical experience in the analysis conducted in tough conditions
- Too many new people on the team, not enough experienced team members who could guide the work

How can people fight for their vision of building the system in the project if they don't have a clear understanding of the system themselves, don't have an opinion, and thus can't convince the client? Problems like these only exacerbate the situation because they entail changes in the plans of individual project members! If the scope changes, its understanding changes, the assumptions change, the client pushes for a different interpretation of a feature, all this eventually ends up on the desks of the programmers who will have to replan their work.

Processes like these are a daily occurrence in any project. As the scale of each minor delay at an individual level is ignored from the project manager's perspective, we could say that everything is happening outside the range of his radar and is not even noticed. Who cares about one or two extra man-days on some minor feature, right? The problem is, however, that over a month, two, or a quarter, it leads to delays that are significant from the client's perspective, and these delays are hard to explain as nobody registered such seemingly unimportant changes. In a sense, the people in charge of the project feel helpless. Each time the project manager asks people what happened, they say that initial plans were wishful thinking. The new planning will definitely be accurate. A quarter passes and the situation repeats itself.

The Source of the Problem

Programmers, architects, developers, and testers are not project managers. They are not experienced in creating schedules. It's just not their domain. They want to focus on their work, but they also forget that management is also necessary and is not reserved for the lazy ones who understand nothing and therefore become managers. Sadly, most companies don't attach emphasis to training specialists in the planning basics and monitoring progress, or how their work affects others and the project schedule. What can we expect as a result? If educating people in making accurate estimates is neglected, the quality of the estimates will be questionable.

The management rarely communicates high-level business goals. They don't inform people early enough (if ever) about important milestones! The people don't usually understand the context of the project or its important aspects. For this reason, they are living in their own microcosm which is very different from the reality of the client and the management. Their priorities are drastically different from those of the project. They have a different feeling of time, pressure, or its lack.

If we don't explain to people what's at stake, the purpose of their work, and its impact on the business, we cannot expect that they will be overly excited or motivated to act.

It's often the case that specialists are not reminded about why their estimates are so important, that they must meet their deadlines at all costs and at all times. That's why people think that their tasks are negligible, unimportant. If they're late, nothing will happen. In fact, a project is a sum of all these seemingly unimportant tasks! If everybody approaches their commitments in this way, the project will be constantly delayed and uncontrollable.

Another important issue is taking ownership for finishing one's tasks. Specialists often assume that it's the managers' job to solve all problems. Thus, when faced with problems, they either don't report them or expect that the manager will handle everything. One could give a lot of examples of such behavior. Mistakes which are pushed on other programmers and wait for God knows what. Blocked work, which is not done as somebody is to blame and first needs to deliver the prerequisites for this person to start – for now it's only waiting passively. Such behavior is notorious and it's one of the main reasons for delays. It could even be called applied negligence.

People need to be explained that each person in the company, regardless of their role, must take ownership for finishing their tasks. They must fight for it instead of delegating them to someone else and waiting. They shouldn't make excuses claiming that it's not their job. Everyone must fight all the time! Take ownership and deliver!

The pitfall of planning is that people don't take responsibility for their own estimates. It's rare to find a person who feels guilty about not delivering something on the day that they declared themselves. It's easy for them to give various excuses, most often in the vein of: "it turned out that it was more work." Unfortunately, it's a serious problem for the organization. No plan can be kept under such conditions. In a perfect world, each person should do everything and anything to deliver work on time that they indicated or to estimate more realistically! The careless attitude is a curse for any project and organization. When such behavior is tolerated, the company becomes uncontrollable in a short time. How can we make plans based on estimates that nobody respects and nobody fights to meet the deadlines? The manager is accountable for the performance of his business unit, but his control over them is indirect. They usually manage team leaders and it's they who are line managers for people doing the actual work.

It's the specialists who have control over the work pace and meeting deadlines because they do the same work themselves! That's why the key to success is working directly with them and their line managers.

Of course, we often experience something that is called "pushmittent" (a combination of the words *push* and *commitment*) when the management asks for accurate and realistic estimates, but then employs various means of exerting pressure, in a way forces people to shorten the time (as was described in the previous chapters, especially in the context of sales). Then it's only a matter of time when these estimates will change and become more realistic (tasks will be delayed). It's not surprising.

Building Solid Foundations for Planning

The foundations of responsible planning are low-level estimates – estimates of how long individual tasks will take. In other words, the basis for a realistic schedule are the estimates provided by specialists – programmers, architects, analysts, testers, developers, that is by the people who will perform certain tasks. Thus, it's of major importance to invest in educating these people because they're the ones who will do the work and have direct control over it. It's necessary to instruct them how to plan, which tools to use, explain to them various, tested estimation techniques, ensure that they are coached by more experienced colleagues who know exactly what works in this particular organization and what doesn't.

It's necessary to create a sense of ownership, especially among specialists. Shaping the right attitudes is key. I estimated the task, I suggested the delivery date, so I am the one who needs to keep my word! I care about doing my job! I don't push the work over to other colleagues. If my tasks depends on my colleagues' tasks, I actively urge them to help me. I proactively help, take initiative instead of waiting and wasting precious time. Without this sort of attitude, it's pointless to hope for meeting the deadlines and success in the project. Everybody must have the same goal in mind.

It's necessary to create a comprehensive, holistic understanding of the project in the team. It's necessary to explain the client's problems and needs, how they should be addressed, which functional aspects of the system are crucial, what the main milestones are in the schedule, what exactly needs to be achieved, why it's important, etc. Moreover, it's not enough to mention it from time to time. It's necessary to talk about it over and over again!

People must understand and feel that their work is very important, that it depends on them whether the project succeeds or fails. They're not just numbers in the organization, it's the sum of their tasks that combines into the outcomes both in terms of the project and the entire organization!

Commitment is something that is built from the lowest organizational level. It's not enough to motivate the executives and department managers. It's not enough to exert pressure on the top or middle management. These are not the people who do the actual work. There's no guarantee that they transfer the motivation and excitement on those who have real influence on the outcomes.

The problem must be solved in the systemic way. It's necessary to make sure that the messages are conveyed to each person in the company, and especially to the most important people: programmers, analysts, architects, developers, and testers. These are the ones who have most impact on the project, nobody else! And it's they who have to be motivated in order to deliver!

It's key to consistently build high self-esteem among employees on each level of the organization. It's not true that the key to the company's success are the highly remunerated executives. In reality, the most important elements are specialists and team leaders who work with them! In most companies, these people feel insignificant, that they're unimportant, that nothing depends on them, that they're easily replaceable. It's absurd, and a lot must be done to change this situation. If people feel their worth, they know that they're cared for by the employer, that they're important, that everything depends on them, they'll work with devotion and zeal. And that's exactly what we're aiming for. The company should also design the appropriate bonus systems which show who is seen as an asset. It's not enough to say "you're important for us" if the whole splendor and financial gratification then goes to a narrow group of selected people.

It's the team leaders (and not the executives) who act on the frontlines. They are the ones who directly manage the people doing the actual work. Thus, they're the ones with the biggest impact on the project plan. They should monitor progress on a regular basis, identify the risks, and understand their influence on the specialists' tasks, help in solving problems, actively seek solutions. Team leaders directly affect the team's motivation, whether team members understand the goals, what's at stake, the meaning and purpose of the work. Therefore, instead of organizing prestigious and costly programs such as an MBA for the top management, it's a better idea to invest in training for team leaders! They are the ones who should be taught leadership approaches, communication, motivation, planning techniques, methods of controlling progress, budgeting, reporting, etc. Thanks to competent and well-educated team leaders, we increase the chances that specialists doing the actual work are motivated, understand the goals and priorities, that there's someone who will help them in planning the work, teach them to be responsible, and to meet own deadlines.

We often forget that for programmers, analysts, developers, testers, it's their line manager, the team leader, who is the face of the company, representing all its values, standards, virtues, and vices. If we neglect team leaders' development, in a way we agree on low quality on a large scale, regardless of whether we hire experienced MBA graduates, the organization will not work well, planning will be inaccurate, projects will always be delayed, and most staff will perceive the company as chaotic and unprofessional. Why is it most often assumed that the solution to this problem is a so-called new broom, meaning a new top executive? The answer is incredibly simple. For top management, it's much faster, simpler, and cheaper. We sign a contract with a new person and we're done – that's it. It's much tougher to craft a long-term training program for the entire company, launch it, monitor its effectiveness, and adapt it. It takes much more effort, money, and time, it requires constant attention of the top management, it's tedious, and the results are distant and difficult to measure.

Of course, the point is not to hire just anyone for an executive position. In this case, it's also important to assess their knowledge, competencies, and experience. The fact is, however, that this problem is better diagnosed and obvious in some way. Whereas investing in the education of specialists and team leaders is often overlooked. And certainly it has no high priority attached. The management isn't aware of the huge impact that the ability to plan and monitor progress by specialists and team leaders has on the performance of the entire organization. These people are just a great majority of employees. They are the core of the business. The scale of the problem is therefore much bigger and that is why it has to be focused on to get better outcomes and improved performance.

Summary

- Low-level estimates are the foundation of planning. Accurate estimates of tasks translate into a stable, overall schedule of the project.
- The actual work is done at the level of specialists such as programmers, analysts, architects, testers, developers. These are the ones who have direct control over the deadlines.
- Success of the project is, to a large extent, dependent on how each individual plans their work, if they meet their own deadlines, if they take ownership of their own part of work, and if and how motivated they are.
- Most managerial work happens at the intersection of specialists (the team) and the team leader. For this reason, it's important to put emphasis on developing team leaders.

CHAPTER

18

Fixing Bugs – Planning Doomed to Fail

The greatest enemy of knowledge is not ignorance, it is the illusion of knowledge.

—Stephen Hawking

An interesting issue that we face in IT projects is planning how long it will take to fix the bugs reported either after internal tests or user acceptance tests. Most managers, especially executives, on the client's or supplier's side expect a specific date (mostly related to a milestone) by which all identified issues will be fixed. They assume that the process of achieving a desired quality state is no different from any other work, that it can be planned.

Project managers then face a serious challenge when they try to explain that it's a completely different situation, less predictable process, and they are not able to accurately estimate the time needed to fix identified bugs. They strive to explain that fixing current bugs is not all, that there can be new ones, unidentified yet and that it's a completely normal phenomenon, that a lot of

M. Dąbrowski, *Managing IT Projects*,
https://doi.org/10.1007/978-1-4842-9243-3_18

things could not have been anticipated. Unfortunately, these answers aren't usually understood, they annoy the management both internally and on the client's side. Of course, you can and you should work on developing quality assurance processes, improve them and consequently start tests from a much higher level altogether. The problem can be minimized, but it's an extremely rare situation that a delivered system is completely flawless, and the client accepts it on the spot.

The conviction that bug fixing can be planned, that it's possible to plan a specific timebox for that in the schedule, is just wrong!

It's impossible to predict a specific number of bugs. It's not a deterministic process. Certainly you can and you should care about quality to lower the chances of bugs appearing in the UATs. However, you must remember that even planning those tests is difficult for many reasons. You can't predict what bugs will be identified and how many there will be. And this is what needs to be taken into consideration when estimating the initial cost of improvements. Any bug can differ from others in terms of complexity. What seems trivial may require a substantial effort. There are also bugs which can be seen as "onion-like" and fixing these only leads to finding more and more new bugs. Even after all bugs have been fixed, the client might find new ones, unidentified before.

Due to this, committing to any dates when the bug-fixing process will be completed is extremely irresponsible. It's sheer wishful thinking that has nothing to do with reality. We can spend any time we want planning, we might try to believe in that plan, but it turns out futile anyway. Reality will prove that we were wrong. Sure, there is a simple rule saying that the more effort we put into conducting internal tests, the better will be the development process in terms of quality assurance, the better people will be allocated to the internal quality assurance tests, the fewer bugs there will be in the end and the more accurate our plans will be.

All in all, instead of planning the process of bug fixing, it should be measured ex post; on this basis, a trend can be identified and then the project schedule should be adapted accordingly. You need to develop the right practices and tools, decide what specifically we want to measure and how, in order to be able to draw some conclusions based on the results we get.

What definitely needs to be measured is the pace of identifying defects on the client's side in a given unit of time. Based on that we can estimate the increase in the number of bugs to fix on an ongoing basis. We should monitor the team's efficiency in bug fixing in a given unit of time. This shows us how quickly the number of bugs decreases. After compiling these two datasets we get a trend showing the actual situation. Thanks to this we can draw some general conclusions about the direction in which we're going and when more or less we can expect to achieve the required quality level.

Nevertheless, it's necessary to keep repeating to the client (and own management) that these are only estimates and not milestones in the schedule. They can change on an ongoing basis depending on what we do. Based on the collected data we can make decisions such as: to strengthen the team, to change priorities, to look for the real causes of the quality issues, to optimize. Eventually, it turns out that measuring the process ex post, monitoring the trend in bug fixing brings much better effects and work stability than hopeful planning and trying to cast a spell on reality.

The biggest problem for the supplier is convincing the client to try the preceding approach. There is a schedule in the contract after all. The client has a right to expect a timely delivery, regardless of the circumstances and the adopted project management approach. This is an objective fact. You must remember, however, that the "sacred" schedule was a result of negotiations, pushbacks, and wishful thinking on both sides, expectations as to when the project needs to end. It is precisely this "unchangeable" project plan that is detached from reality and, at the same time, is a source of conflicts between the two parties.

The talks about how it was supposed to be or how it should be, lead nowhere. It's important where we are now, what the situation is and what course of action is possible.

In this context, in an uncertain situation, the only reasonable approach is to monitor, collect the data, identify trends, and react accordingly. It's therefore necessary to persuade the client that casting a spell on reality, unrealistic expectations, putting pressure on the supplier or particular people does not bring any positive outcomes and changes nothing.

It doesn't mean that we shouldn't try to meet the deadlines. Sure that you have to do your best to deliver the project on time. But at the end of the day, it's impossible to change the rules of physics. We cannot inflate the team tenfold, we cannot make the day longer, and we cannot guess how many bugs will be found, what they will be, and how much time will be needed to fix them.

Summary

- It's a common mistake to try to determine in advance (at the beginning of the project) how much time will be needed to fix bugs and then to make the team accountable for meeting these unrealistic, abstract commitments.
- Instead, you should create the right quality measurement processes for the developed products, and those processes will provide real data at regular intervals. Thanks to this, it will be possible to react and adapt on an ongoing basis.

- This approach is not the same as ignoring the schedule. The thing is to avoid false assumptions, which only lead to problems, disappointments, and conflicts between the parties.
- The client should be persuaded from the beginning that instead of rigidly following unrealistic plans, it's better to work with numbers and data and deliver more quickly as a result.

CHAPTER

19

Managing Delays

Delay is preferable to error.

—Thomas Jefferson

It's more than certain that delays will happen at some stage of the project. Shorter or longer, sooner or later, but they will definitely happen. To a large extent, managing a project is organizing the team's work in such a way that blockers are avoided or minimized if they do happen, in order to create an environment that will enable the team to work effectively. Delays happen and they always will, especially at the level of single, individual tasks done by specific people. It's a consequence of the following:

- It's impossible to precisely foresee the future and also how long a specific task will take in reality.
- It's impossible to predict what kind of problems a specific person will face, how complex they will be, and how long it will take to solve them.
- Nobody can anticipate what cooperation will be like with other people on the supplier's and client's side.

M. Dąbrowski, *Managing IT Projects*,
https://doi.org/10.1007/978-1-4842-9243-3_19

The preceding list shows how difficult it is to plan work for one specific person, when in fact a project plan is nothing else but combining work plans of individual people, then teams, and finally organizations. Project planning is also based on company-level processes such as recruitment and acquiring employees, purchasing, contracting subcontractors, deliveries, and many more.

It's rather obvious that planning work in a way that any delay will be avoided (in relation to the initial schedule) is plainly impossible. Yet it doesn't mean that a project can't be finished on time. Anyway, even the very definition of finishing on time is incredibly misleading. Planning is based on the initial assumptions which turn out to be wrong during the project. For example, a real scope of work is larger than it seemed in the beginning and the time for doing the tasks is just longer. Some problems turn out to be clearly more difficult.

Project management should focus on achieving the client's business goals in an objectively optimal time frame, rather than on minimizing delays in relation to the initial schedule which is most often unrealistic. If the supplier is fully dedicated to the client and their business goals, delays are no longer a problem.

Who Is to Blame

If you look at it from the right perspective, you can see that projects follow a certain pattern. Moreover, you could even try to create a generic list of potential reasons for delays.

On the client's side, they are

- Problems with finding people with the right competencies who also have the necessary project-related domain knowledge and should take ownership for a particular part of scope
- Problems with these individuals' availability (taking them away from their daily tasks), which results in continuously extending the time for completing tasks
- Difficulties with delivery of the input materials necessary for the supplier to start work (data, analyses, requirement documents, specifications)
- Organizational problems when several departments on the client's side take part in the project and only one of them tries to coordinate all involved parties
- Lack of decision-making power, which results in looking for the persons responsible for a particular problem every time it happens, too long discussions, management chaos

- Delays in delivering materials (hardware, test environments)
- Lack of control over other subcontractors delivering other parts of the scope

The typical reasons for delays on the supplier's side are

- Lack of domain experts in the initial phase of the project
- Lack of resources in the development departments at a later stage
- A different state of the sold products than was declared to the client, which results in extra time needed for development
- Slower than expected pace of delivery (a weaker team, higher complexity of problems to solve)
- Low quality of deliveries, resulting from poor organizational culture on the supplier's side or from too fast pace (too ambitious) of delivery

The worst situation for the supplier is when the fault is only on their side. The best situation, on the other hand, is the opposite one, when the difficulties on the client's side give the supplier more time to handle their own problems. In reality, the situation at hand is usually complicated when delays are both on the client's and on the supplier's side and explaining the cause and effect relationship is hard – to say the least.

Understanding the sources of project delays is important for many reasons, for example, it allows to document the course of events properly, provides a possibility to present the status of the project in the steering committee meetings in a positive light, and to create better perception of the project in the client's eyes.

Shaping the Client's Perception

A basic mistake is assuming that the client (on every management level) is fully informed about the real status of the project. Unfortunately, it often turns out to be just the opposite. Decision makers are receiving the supposedly objective progress of the project. It happens in official meetings such as the steering committees, but also during informal interactions both internal and with the supplier's representatives.

The thing that is more important than the real reasons for delays is the client's perception, the way in which he is looking at the current status and the chances for a partial or full success. If the supplier is seen as a partner who is trying to help, their faults are usually tolerated. It is similar when during the steering committee the project team on the client's side defends the supplier claiming that the complications which happened are objective and the supplier has done everything that they could to prevent them.

The executives (client) most often pay attention to the following aspects:

- The way that both teams cooperate (whether there is some friction or conflicts, whether people want to work together, if they speak "in unison" in meetings)
- If the project team on the client's side still believes in the project's success and in the supplier
- If the project's business case is still up-to-date
- What the supplier's attitude is in the face of problems (whether they act in a constructive way or they blame the client, cheat, and lie)

As you can see, the management still attaches a lot of attention to the soft aspects, in particular the cooperation between both sides. That's why, on the one hand, the project manager must keep maintaining a professional attitude, but on the other, he must also deliberately shape the perception of the project by the client. The project team on the supplier's side should also take care of that. Everyone has their counterpart on the client's side. It's an oversimplification, but if most decision-making people on the client's side are convinced that the supplier is doing everything to solve problems, the project is highly likely to succeed.

A typical example of a steering committee in IT projects is as follows. You start with informing about the delay whose causes are difficult to identify. The complications happened on both the client's and the supplier's side. The project manager has two possibilities: he can try to lay all the blame on the client and persuade the management to believe in their version or present the situation in a more objective, impersonal way, focused on the solution and sharing the narration that was previously developed with their counterpart on the client's side. In most cases, it's a good idea to take the cooperative stance, thus demonstrating that both teams can solve problems together. Challenges and delays happen and will happen. The management is fully aware of that. This is why a lot of attention is paid to the quality of cooperation and the attitude in the face of adversity. Of course, depending on the geographical region and the national culture, the communication style might slightly differ, but what counts is still the intention, showing that the client's interest and their people are important for the supplier, that constructive cooperation matters to him.

You need to remember that projects are conducted for money and there is always someone who looks at the outcomes from this perspective. On the client's side, it's important to achieve certain business goals (savings, higher effectiveness of business processes, increasing the number of clients in a shorter time). The supplier wants to get references and issue invoices. Thus, delays are objectively harmful. Even if the perception of the project is very positive, the client is happy, notorious problems and delays may force the management (on both sides) to stop the project.

Documenting the Causes of Delay

If the delay happens, you need to remember to generate as many documented artifacts as possible: emails, slide decks, meeting minutes and written agreements, etc. In such situations, the risk of stopping the project by the client has to be taken very seriously, so all the data related to the causes of the delay must be collected. A particularly important source of information (from the point of view of the future court case) are the emails from other team members. Even the slightest faults on the client's side may matter in the future. Everything that is or might be related to the delay has its weight. What is more, even the circumstances (on the client's side) that are not directly related to the delay are important as they may potentially be a part of an argument to prove their indirect influence on the problems on the supplier's side.

If it ever happens that the client stops the project and terminates the contract, everything that happens afterwards (exchanging demands, negotiations, court case) will be largely based on the materials gathered by the project manager!

If the project documentation is convincing, it will help in creating a history of events in the project speaking in favor of the supplier, the supplier can feel safe. The judge and the court expert always assess the project based on the documented materials. Later on, you can present any (real or imaginary) version of what happened, but the winning party is usually the one which can support their interpretation with a bunch of documented "facts."

Accepting Delays

The simplest strategy in the face of delays is just accepting them. At first glance, it might seem like avoiding the problem and the stress related to it. Contrary to what you might think, accepting delays is a deliberate approach which makes sense in some circumstances. It's usually the first strategy of choice, especially in the beginning of the project when

- First, insignificant delays occur
- Neither of the sides was aware of the problems both on their side and those resulting from cooperation

- Business goals remain the same
- Belief in the team and its strategy of handling the project is strong
- There haven't been any important or unsolvable problems calling for some radical actions
- There are some objective initial problems, for example, the extended time to form teams

In most cases, initial delays occur for objective reasons, for example, they result from the characteristics of the operations at the client's (and supplier's) company. They are not actions that can be blamed on the team members. They are not management mistakes, either. They are not the consequence of the lack of willingness to work or increasingly challenging technical tasks. The project manager can expect some organizational problems on the client's side and use the "extra time" to fill some gaps on the supplier's side.

Changing Priorities and Rescheduling the Project

Another strategy used in case of recurring delays is changing priorities and adapting the schedule accordingly. It is used in the middle phase of the project when

- The objective technical and organizational problems are well known (internal and those resulting from the external environment)
- Both the client and the supplier have identified bottlenecks at their respective ends
- It's obvious what is not working in cooperation between the two sides
- The initial business goals have partially lost their appeal, they are evolving and the way to execute the project needs to be modified accordingly
- It may be decided at this point that dealing with a specific area is easier (or more difficult) than another and bring much higher (or lower) value

It often turns out that the project can bring more value to the business at a lower cost, that it can be completed much sooner. It's enough to switch the order of some stages, optimize the scope of particular deliverables. The difficult part is to analyze what works and what doesn't, reconsider current business goals, project capabilities of both the client and the supplier, and to find the most optimal trade-offs between these elements.

One of the clients was dealing with a problem of obsolete IT tools. The main system was so old that there was nobody at the company who took part in designing it. Nobody knew how to develop it knowing that new business requirements are underway. That's why an external subcontractor was selected and their task was to implement their proprietary product.

After over a year into the project, it turned out that the client systematically forced the supplier to copy the AS-IS systems and business processes into their product, which effectively meant rewriting the old system (with all its deficiencies) from scratch , and that is a very costly and time-consuming thing to do. At the same time, the schedule imposed by the client's IT department made no sense as the planned order of delivering particular modules was contrary to the client's business strategy. In other words, the entire effort made by the supplier was focused on the elements which were of negligible value to the client. The solution to the problem turned out to be the change of priorities and the right replanning of the project so that the most crucial (from the business point of view) functional modules of the new product were implemented first.

Reducing the Scope

The third interesting option that a project manager can go for is reducing the scope of the project. It's not graceful for many reasons. Young management trainees are always taught that the most important aspect of project management is to deliver the planned scope. The thing is, however, that in practice, there is no such thing as an objective and indisputable scope. Just like in quantum physics, the interpretation of the project scope depends on who is looking at it, so on the particular stakeholder. It is very common that at the client' s organization there is a conflict between the management – business – IT – users. From this perspective, reducing the scope can't be about removing a random part of the work, but about identifying the elements of the scope which had been wrongly (based on some subjective wishes) deemed important. In order to make reducing the scope possible, it's necessary to get the support of the key persons at the client's organization, as they are the ones who need to be convinced to make difficult decisions and to fight against opponents within their own organization.

Reducing the scope is the best strategy when

- At a particular stage of the project (it may also be true for the entire project), there are constant delays, and deliveries are postponed all the time
- Both parties feel that the planned scope is too ambitious and too extensive, and a large part of it is just wishful thinking
- The business goal can be achieved with a much smaller and pragmatic effort
- The time of delivery is important (for various reasons the date is set and cannot be changed)

The most common model of projects in IT is replacing a legacy system (or systems) with a new one. It's usually connected with changing the way in which the system is used by end users. The main and daunting task for the management is to deal with the users' habits. It's almost always that they try to enforce solutions which resemble the old ones. Their aim is to maintain the status quo. It results in

–An unending stream of changes in the project scope

–Distortion of the project's business goals

–Creating a monstrous system in place of an elegant product

–Increasing costs both for the client and the supplier

–Extending the schedule

To a large extent, it is the client's management's job to lead the project stakeholders. That's the area where we have the possibility (and maybe even the necessity) to keep reducing the scope – bringing the project back on the right track.

Splitting the Stages and Extending the Project

Clients often set audacious goals. At the stage of drafting the strategy, preparing documents and presentation for the executives, everything seems doable. As the saying goes: "Anything goes on paper." The problem is that business strategies are crafted by managers responsible for a specific line of business and they hardly ever ask for advice from the specialists who will then be accountable for the implementation. Eventually, it turns out that the changes cannot be made as planned due to some technical and organizational constraints. In such projects, you need to demonstrate very clearly that in the proposed shape, the undertaking will end up in a disaster. Then comes a time to divide the scope into stages, change the parallel actions into a sequence, and extend the project in time.

The splitting strategy is used

- Most often in large transformation projects
- After the first deliveries, when it's clear that on both sides there aren't enough resources to carry out too many tasks at once
- When it turns out that delivering everything at once (the so-called Big Bang approach) won't work
- When it becomes evident what both sides can effectively complete in a certain time
- When the client realizes that the initial strategy was overly optimistic and objectively unrealistic, both in terms of the supplier's capacity and the client's organizational capability (it would be impossible to consume the delivered products anyway, to transform the organization and business processes, so it wouldn't be possible to use what was delivered)
- When, after completing a certain phase, a large part of the team on both sides focuses on maintaining (of what has just been delivered) and focus on the following stages naturally decreases
- When both sides are unable to work on several project stages at once and need to extend the project and deal with the stages in a sequential way, one after the other

Both suppliers and clients (especially them) need time to understand that the initial project strategy did not make sense, that it's not really about delays but about inappropriate and too ambitious approach to the project. It often happens that at the sales stage, suppliers approve the unrealistic plans because otherwise they wouldn't be able to start the project. They know, however, that at some point, the client will realize that a project like the one at hand is in fact conducted at a different cost and in a different time frame than was initially assumed.

Summary

- It's certain that delays will occur sooner or later.
- Usually the reasons for delays are on both sides, which needs to be meticulously documented.
- The way the client looks at the delay and the supplier's reaction are important and that's why the client's perception needs to be proactively shaped.
- The simplest delay management methods are the most effective ones.

CHAPTER

20

The Whole Truth About User Acceptance Tests

The best is the enemy of good.

—Voltaire

You could imagine a perfect situation when there's no time pressure from the client, there is no emphasis on annual performance and invoicing, and the whole project can be executed by the book, following the best practices of the IT trade. Then we would have time to develop the commissioned system so that it fulfills all the client's wishes and to perform all the work related to achieving the desired quality. We would probably spend a lot of time adjusting the system's performance under real-life load conditions, deliver the non-functional requirements, etc. In theory, approval procedures of such a system

M. Dąbrowski, *Managing IT Projects*,
https://doi.org/10.1007/978-1-4842-9243-3_20

would be just a formality. The client would conduct user acceptance tests quickly, there wouldn't be any major bugs, after that the system could go straight to production.

Sadly, the reality of IT projects is quite different. We're always faced with a lack of time and resources. The works are usually delayed and the delay is significant. The client wants to get the deliverables as soon as possible, as he is losing profits and can't reach his business goals on time. Internally, there is pressure on performance and invoicing. The company must make profits and pay salaries. The project team makes desperate attempts at delivering faster by performing multiple tasks at the same time, often doing things that are not entirely sane. Components which should be developed sequentially are developed simultaneously, which results in an increased number of integration and quality issues, and eventually it is necessary to refactor a big portion of the code.

In practice, IT projects are a constant fight between quality and time. On the one hand, you have to issue invoices as soon as possible; on the other hand, it's necessary to ensure the right quality of the system and meet the acceptance criteria. In the end, the pressure to deliver both on the client's side and the supplier's side leads to a situation where we start user acceptance tests too soon, testing a system that is not ready for that by any means. First of all, it hasn't been internally tested, and besides, some of its features are still being developed. As you know, it results in much longer testing periods and reporting a large number of bugs. Clients usually even say that they are treated as alpha testers. The supplier works in three main areas: helps the client with testing, is hastily trying to finish the development of the missing features, and on top of that runs internal quality tests.

As you can guess, it's walking on thin ice. In theory, at first glance, it's clear that it's a suboptimal way of working. Most people think that they could easily find a solution to this problem, or even that the management is incompetent and has no understanding of the IT domain. What's more, these opinions to some extent are justified – "to some extent" being the key phrase.

Why User Acceptance Tests Should Begin As Soon As Possible

The constant pressure, lack of time, working according to crazy schedules. At first glance, it seems that the best solution would be to delay the project and do all the tasks in the right order – with no rush, peacefully, thoroughly, precisely, and in a way that is described in coursebooks. But is that really so? Would that really be the best solution? Would it be safer for the business? Will the client's business wait? Will they respect the rules of the IT domain? Does it matter at all? Business is not controlled by the IT guys only. People don't always use logic in making decisions.

Most often, both from the supplier's and the client's perspectives as well as for the sake of the project itself, it's better to start user accepting tests sooner than later, even if the system is not complete, even if it hasn't met the quality criteria and even if the system's verification and approval process will be significantly extended.

In large transformation projects, it's often the case that the client must wait for months or even years to get the first tangible deliverables. With time, people become less and less patient. If we add to that constant delays, we get an explosive mixture of uncertainty, distrust, disbelief in the supplier, and even doubts related to the purpose of the project itself. After some time, the client needs some form of evidence that the transformation makes sense. Usually it will be some part of the working system which shows to the future users, the management, the business that there are real, concrete outcomes of the project.

Therefore, both for the client's and the supplier's organizations, it's much safer to start the process of acceptance testing even if it's going to last for weeks or months to come. We can always say that we're in the process of shipping a particular stage of the project. This kind of communication is much better received. Besides, both sides need it. In theory, the phase of user acceptance tests could be postponed due to quality improvements under way. The question is, however, if it's better to delay the start of the shipping process by three months due to quality issues or to start user acceptance tests and conduct them for three months more? The client's management may not be patient enough to keep waiting.

Initiating the process of user acceptance tests is a big success for both sides of the project! Both teams feel that they're delivering after all, that they can complete something.

Moreover, future users begin to familiarize themselves with the features of the new system, start to live it, get used to it. If they can "play" with the new tool which is usually more technologically advanced and has more features, they will be more willing to wait, accept issues or delays. The project management on the client's side also reports to someone. When they have evidence in the form of user acceptance tests, they can prove that there is progress in the transformation they're accountable for. Without tangible outcomes they can have problems to justify the budget for financing the project in the future.

Starting user acceptance tests also helps to manage the scope. There's always a degree of freedom in interpreting what should be developed. You can't totally eliminate users' individual expectations. The bigger the project, the more difficult it is to manage. Entering the shipping process entails complete feature freeze. In theory, no changes should be made in production. Obviously, life is not black and white: clients usually try to enforce changes in the system until the last moment. However, in the shipping process, it's much easier for the supplier to escalate situations like this. And the client's management tends to take the supplier's side in those instances. Thus, if we waited for two months for the shipping process to begin (due to insufficient quality), the client could still demand product modifications. If we initiate the tests sooner, even if they last two months longer, managing the stream of incoming requirements will become a much easier task.

From the supplier's point of view, invoicing is very important. His business is based on delivering projects after all. They're a source of income. Most often, user acceptance tests are an incredibly important milestone in the payment schedule. It means that the features planned for a particular stage have already been developed, so the supplier has borne most of the costs for a given project phase. That's why starting user acceptance tests is invoiced and the amount on the invoice should cover the supplier's costs.

In other words, the amount of revenue connected with the initiation of the system shipping is usually substantial. The date of issuing an invoice matters for all businesses. It translates into quarterly, semi-annual, and annual bottom lines. From this perspective, there is a huge difference between issuing invoices in December and in January! Organizations can't function detached from their finances. If you start UAT in December this year's bottom line is saved, whereas if you start the tests a month later, the department's performance decreases significantly. How are you going to explain to the management that the revenue of several million wasn't invoiced in a given year because in your view the quality criteria were not met? It's enough to start the testing process sooner and continue for a month longer. To make it even funnier, big corporations operate on yearly budgets. And thus the clients themselves want to pay for particular milestones in a given calendar year and don't want to delay the payments so that they can keep their budget for a particular project. So, they will also insist on starting the UAT sooner.

Depending on the agreed payment schedule, the client has the right to withdraw from the project without incurring any significant costs. The period from starting the project to the first tangible deliverable is therefore filled with uncertainty as to the future of the undertaking. The readiness to ship the system or a part of it and invoicing the significant amount of money related to this milestone seriously impacts the client's engagement in the project. If they decide to pay for the readiness for UAT, they make a conscious decision to continue the cooperation with the supplier.

Each amount paid by the client reduces the risk of ceasing the project! The higher the investment, the more difficult it gets to withdraw!

You can't forget that initiating the user acceptance tests or the readiness to conduct them is an important event in the project. For this reason, as early as the sales stage, while the contract conditions are being discussed, clients protect themselves by introducing high penalties for delays in the readiness to ship the system or a part of it. Thus, it's clear that it's much safer for the supplier to start UAT sooner, even if the system is not ready for that, and avoid penalties, and then finish the tests with a delay for obvious reasons.

Almost every large IT project will be a proof for these observations. For example, we could take an average from the transformations in three companies representing three different industries. In all of these cases, the supplier was too optimistic in estimating the scope of the project, which resulted in constant delays, escalations, and threats to terminate the contract. Each time clients waited for the first deliverable for so long that the future of the project was very uncertain. The belief in success was incredibly faint. Only after initiation of the UAT by the suppliers, did the situation change, even though the shipping processes were extended from several months to almost a year! Fortunately, after starting the testing process in each of the projects, the client's management could demonstrate progress, the initial success, and the client's organization and future users could play with the new system, they changed their opinion of the supplier, they were convinced that the expected outcomes are very close, and they made allowances for the insufficient quality as the systems were "almost" ready.

From a psychological perspective, this phenomenon is extremely easy to explain: it's better to have something "almost" done and test it for a longer period of time than to wait apprehensively to see if it even exists and if we're not being cheated and misled. Even though it's contrary to any good practices of software development, in real life, you must sometimes make decisions which are not perfect, but the best in a given situation. If the suppliers in question had decided to keep delaying the shipping under the pretense that they were still improving the quality, the clients would have terminated the contracts and improving the quality would have lost any sense. However, the projects were completed and eventually everybody was happy. It might not have been correct from the point of view of theory and coursebooks, but it was beneficial for the suppliers and the clients alike!

How to Plan User Acceptance Tests Pragmatically

You could easily say that there is never a perfect moment to start UAT. There's always a list of internal bugs that should be fixed first. There are features that can be perfected so that they work even more seamlessly, others barely work or there is some friction. There will be parts of the system which are not as stable as we would like. The system is hardly ever ready to perform well in real life conditions.

On the other hand, user acceptance tests tend to last significantly longer than expected. The client will usually test much longer because there are not enough people to conduct the tests, they don't understand the system, or can't use it effectively. There will be endless discussions about whether a particular system behavior is an error or a planned behavior. There will be problems with continuous delivery of new versions (packages, releases, hotfixes, etc.) with bug fixes, most certainly the client won't be ready with data migration, and thus the datasets used in UAT will be incomplete. This will only increase doubt about whether a particular test case doesn't succeed due to the bugs in the system or due to the migration issues.

Finally, even after the user acceptance tests, it will be necessary to wait with starting the system in production. The client will need time for data cleansing and that could take months. The supplier will certainly have to improve the system performance. All this will give you some extra time for developing the missing features, fixing bugs, stabilizing the system, adjusting performance, etc. You could safely assume that if the user acceptance tests were planned to take a month, they will take two or three. This is another reason why initiating the UAT process sooner makes sense even if you're not completely ready.

The rule pertaining to the pragmatic planning of user acceptance tests when the system is not yet complete is as follows:

- The scope of delivery should be divided into closed functional blocks.
- The UAT schedule should be drafted in such a way that the order of testing particular functional blocks aligns with the level of their readiness and quality.
- Development should be focused on improving the missing blocks which have yet to be delivered during the UAT tests.

To cut a long story short, instead of waiting for the start of testing until the system is complete, it's better to start user acceptance tests with the features that work properly and, in the meantime, finalize the development of the missing features and fix some quality issues. Testing of particular blocks will take longer anyway, which will give you time to improve the areas that will be tested at a later stage.

Clients also care about time, they don't want to waste it and that's why most often they treat this issue pragmatically and accept this approach. You can easily protect yourself by adding full regression testing phase at the end (formal testing of all test cases) which usually takes a limited time as most issues had already been discussed, the client is familiar with the system and knows how to test it effectively.

Of course, there are industries and situations in which the preceding procedure is absolutely unacceptable due to the very restrictive approach to quality control. Nevertheless, in most cases, we should try to start the UAT process as soon as possible.

Looking After the Testing Client

You can't accept a situation in which the client is testing with no support from the supplier. The future users hardly ever understand the features of the new system. That's why leaving them to their own devices will only lead to a situation in which more bugs will be reported, and in reality, they will not be faults but the result of lack of understanding of the system's functionality!

A good practice is to conduct training on the use of the new tool before the start of user acceptance tests. Besides, experienced trainers or analysts should be sent to the client so that they are on-site throughout the testing process. They should proactively monitor the situation, check how the tests are going and – if necessary – explain how the system should be operated.

In order for the acceptance tests to be effective, both sides should appoint coordinators. User acceptance tests are not mindless clicking in the system. There is a specific number of test cases that should be performed. There are usually quite a lot of them. You need discipline and planning, task distribution, progress monitoring, etc. Moreover, test managers should sum up the results at the end of every day and decide about the priority of particular bugs to avoid endless discussions between users and developers. Bugs can also be weighed. However, for the users, all bugs are critical. Working in this way, without priorities, is a waste of time. A business pragmatic point of view isn't usually aligned with the users' perspective. That's why you need to make sure that the reported bugs are also verified by business representatives who can make responsible and pragmatic decisions. Thanks to that, we make the testing process more efficient, bug fixing quicker, and as a result, the system will be ready for shipping earlier.

Formal Acceptance Criteria – Theory and Practice

Each contract specifies formal acceptance criteria for the system. They're usually very strict such as: zero major bugs and up to five medium class bugs or performing all test cases with a positive result. The problem is, however, that these parameters are almost never achieved. Moreover, even months after being in production, the system does not reach this level of quality. Of course, both clients and suppliers know that.

That's why you should try to convince the client to a more pragmatic approach and toward the end of the UAT process make a list of all bugs and classify them as

- Minor bugs
- Bugs which need to be fixed by the end of user acceptance tests
- Bugs which need to be fixed before the cutover, which can take up to ten-odd days
- Bugs which need to be fixed before the system goes to production
- Bugs which can be fixed after handing the system to the users, with a set date by which they need to be fixed

Awaiting an ideal situation, meeting all the formal acceptance criteria can last forever. The larger the scope, the longer you have to wait. In business, we can hardly ever afford to waste time in the name of principles or theoretical rules. There is no comfort in making obvious decisions which are simple, just right, without any side effects. There are better and worse choices. When both sides care about fast progress the best approach is to act like it was described previously. Why block shipping the system if the reported bugs are acceptable and they can be fixed within a few months after starting the system? The client could potentially execute all formal contract agreements, but what for? What is the benefit of that?

The main problem in persuading the client to adopt the described approach is people. The further they are from the business and real demands, the more dogmatic opponents we have and the more futile the discussions are. Test leaders, for example, will try to classify each minor mistake as the one blocking finalization of the tests, but if we take the list of bugs to a manager accountable for a particular line of business or business unit, he will be more pragmatic. They will be able to make conscious decisions, choose which errors they can live with as a department, what needs to be fixed and by which date. That's why the proposed solution must be discussed with people who are really accountable for implementing the system (systems), rather than with those who don't have the decision-making power or with third-party companies or subcontractors who need to act as "bad cops," which means they look for problems and escalate them instead of solving them.

Summary

- In most cases, initiating the user acceptance tests should be considered as soon as possible even if the system is not completely ready. It's not a perfect decision, but it's pragmatic and most often beneficial for the client.

- The fact of announcing the readiness of the system for user acceptance tests solves a lot of problems both on the client's and on the supplier's side. Both sides can celebrate success. The belief in the project rises in both organizations. The client's management can present tangible outcomes and save the budget for the project. The supplier can issue an invoice and cover the development costs as well as avoid penalties. Users can get familiar with the new tool. The fact that the system is not ready and the tests will last longer gets less important.
- Even if not all features of the system are complete, initiating user acceptance tests still makes sense. In practice, they last longer anyway; they can be divided into modules, we can plan the testing schedule accordingly, and at the same time develop the missing features and fix quality issues.
- Before starting acceptance tests, it's good to train the client on the delivered system so that the reported issues are not a consequence of the lack of understanding how the system works. The client should also get support during the tests.
- It's rare that all formal acceptance criteria are met. That's why you should act pragmatically, apply common sense, and convince the client not to block the system's commercial launch due to some quality issues which can be fixed at a later stage.

CHAPTER

21

If You Want to Save Money, Invest in People

An investment in knowledge pays the best interest.

—Benjamin Franklin

The miser loses twice.

—Folk wisdom

A very interesting phenomenon is how various forms of illusory saving affect different aspects of the project: beginning with defining the right recruitment policy, through establishing salaries and the way of filling key positions or budgeting resources. For example, in order to save money, you might hire only "inexpensive" employees, people who lack experience and will have to be trained for several years. It's due to having an illusion that thanks to this approach the company will keep making a profit. It's putting economy before knowledge and quality. Maybe you're a tough negotiator and block salary increases for certain employees using various financial indicators as excuses

M. Dąbrowski, *Managing IT Projects*,
https://doi.org/10.1007/978-1-4842-9243-3_21

for doing that. Maybe you try to avoid hiring experienced (and more expensive) people from the market who could bring fresh spirit into the project, share their completely different experience, competencies, and skills, and perhaps you use an argument that it will negatively affect the organizational culture, and thus fill key positions with random people.

The consequences of short-term savings include problems which in turn lead to financial losses which exceed the alleged savings by far. You can expect difficulties with creating effective, efficient, predictable, and performing teams. They must be based on experienced people. When your best people are unhappy, they leave the company for financial reasons and are eventually replaced by a lot of new people. The company – instead of cumulating domain knowledge – loses it, instead of buying new technology – keeps using some ancient technological stack. It exacerbates problems even further. Talented employees want challenges, they want to develop, so – quite understandably – they apply to companies which offer them both a good salary and challenging tasks they will be able to learn from.

Conducting projects without experienced people results in a lot of serious mistakes. Analysts are not able to talk to clients as equals, they can't defend their visions (if they have them at all), and consequently they agree to most of the clients' demands and requests, which means that the development costs increase by hundreds or even thousands of man-days. Inexperienced project managers are also learning their trade, so they're making relatively more mistakes, they are less assertive and confident, which exacerbates both the chaos in relations with the client and within their own organization.

In short, you could say that the near-sighted "saving" policy leads to losses which far exceed the supposedly saved amounts. But, of course, it's difficult to measure precisely. On the other hand, staff salaries are easy to control and understand. That's where the problem lies.

If you want to save money, you need to hire the best people, both talented and motivated. They will be the ones to bring their knowledge, experience to the table, they will come up with solutions to problems, craft an effective system architecture, make the client "buy" their vision, and eventually deliver on time. A company is its people! The quality of the staff translates into the company's performance! They must feel appreciated in every aspect, also the financial one, they also need to be provided with the right environment to foster their focus on the tasks at hand.

Project Analysis and Client Cooperation

A key aspect of every project is specifying the requirements related to the system to be developed, or the scope of changes to the existing one. It's an incredibly important period during which every mistake, every wrong assumption has far-reaching consequences. If the client's expectations are misunderstood, we'll have to waste time and money later to introduce the necessary adjustments. If we allow the client to freely interpret the scope it may turn out that the project will take much longer to complete, and its cost will be so high that at some point carrying on work will be economically unjustified.

At the start of the project, it's key to acquire people with experience, domain expertise, and such who feel at ease talking to clients, assertive people with a vision of the future system who can convincingly talk about it, oppose the client's lofty visions, push against unrealistic demands, and most importantly, defend the product path so that the current implementation remains coherent with the product development strategy. That's why if an organization lacks such people, they need to be acquired from the market. It's impossible to substitute years of experience or extensive domain knowledge. Clients and suppliers need employees with a certain level of expertise, especially if the product to be developed is innovative and implemented for the first time. If the project analysis is not carried out by people who are able to distinguish between product requirements, generic and specific needs of this particular client, we might end up developing a fully customized system and our organization will not develop as a result. Therefore, if the project manager asks for two or three experienced analysts, it's a serious mistake to try to save money on their salaries by hiring someone with less knowledge or experience.

We might want to avoid spending several thousand a month, use inexperienced analysts who are learning both their trade and the business domain, but we can be sure that the client will take advantage of the situation and will easily convince them to the extended scope of the project, which will result in a cost higher by 1000 man-days. It's easy to calculate which hiring decision would have been more beneficial for the company. It's not about offering pay increases to anyone, you need to monitor the level of staff salaries. It's necessary and there's no doubt about it, but not at all cost, not in every situation, and not when it comes to some key roles in the organization.

You need to bear in mind that in projects, especially in large-scale B2B undertakings, there is no such thing as an objective scope. There is merely a subjective interpretation of it. That's why project managers and analysts play such big roles in defining the scope. If a company wants to make some savings the best way forward is to invest in those people!

In a very large project for a client based outside Europe, a supplier was implementing a modern IT platform consisting of multiple products. The scope of work also included a completely new module that was supposed to be developed from scratch, which required acquiring domain knowledge from the market. Unfortunately, it was decided that the analysis would be conducted by employees with absolutely no experience in the technology that the system was going to support.

As a result, the client enforced his vision of the solution which turned out to be more expensive by several thousand man-days than was initially planned. The supplier did save some money on some people's pay but had to bear losses that were way higher. ***The right decision would have been to invest in people, experience, and competencies at the very start of the project.***

Development

When asked what they would have changed if they could execute their project once again, most managers and entrepreneurs say that they would have hired a small team of very experienced experts and have the system developed from the start. Experienced and smart architects will design the system better in every way. It will be more flexible, easier to integrate, easier to configure, the cost of its development will be much lower, the architectural decisions will not have to be corrected over and over again. Smart and talented developers write code much faster, it's clean, requires less refactoring, and it's easier to maintain. They are experienced in particular technologies, they don't make serious mistakes. They often use the most advanced technologies and libraries, thus shortening the delivery time and making it less expensive. Besides, they know what they're capable of. They know how to estimate tasks, how much time they need to solve a particular problem. They're hardly ever wrong, they deliver on time and are reliable. Having a team like that, you can be sure that the project schedule is secure. If people complete their tasks on time, in most cases the project is also delivered on time.

In many software houses, there are cases where a certain module had to be rewritten several times. It is especially true for the elements of the underlying solutions. At each implementation, it turns out that the system architecture does not support some use cases encountered at the new client's and it's necessary to refactor the code. In one IT company, the key module was rewritten six times! It amounted to several thousand man-days. The core reason for that situation was a management mistake, lack of investment in business or technical competencies of the team developing the faulty component. Only people with long-term vision could have predicted how the system would be used in the future and design it in such a way that would make it possible to keep the cost of configuration, maintenance, and development.

We're often faced with a problem when, during the user acceptance tests, the client finds a lot of bugs and is justifiably displeased. Admittedly, some of the bugs are minor and any tester should be able to find them. However, there are also those – and that's the more challenging type – resulting from the fact that the testers don't understand the business domain or how the end users would be interacting with the system. All this adds up to additional bug-fixing iterations, improving the malfunctioning features. Eventually, it generates quite high costs which are completely avoidable. Unfortunately, it's still a common belief that a tester's work is simple and anyone can be a tester. It's completely wrong! Organizations which approach the aspect of hiring quality engineers in this way and try to make savings on these people's salaries incur much higher costs resulting from poor quality of the delivered software. Not to mention that a lot of testing effort can be automated. But in order to do that, you need people with knowledge and experience.

In many cases, the cost of fixing bugs during the development of a new system amounts to 50–100% of the overall development cost. And unsurprisingly, it turns out that it's characteristic for companies which treat the financial parameters (naturally low enough) as their main criterion instead of focusing on the quality of the candidates, their experience, and competencies. They seemingly save on salaries, but then they pay much more in extended schedules, unending bug fixing, and clients' dissatisfaction. It's still not about paying everyone more than they're worth. No company will sustain that. However, you need to be open to acquire talented people and keep the best people in the company, prevent losing competencies and knowledge due to extremely restrictive and rigid compensation policy.

Give a Pay Rise or Let Go?

Clearly, lower operational costs translate into higher profits. Unfortunately, however, too often key people are let go due to the overly rigid compensation policy. Behind every person there's a lot of knowledge and experience, a positive influence on the team, its motivation, efficiency, and relations with clients. It's difficult to put a price tag on all of these. When you let such a person leave the company, you need to hire someone new, seemingly cheaper who will take years to reach the level of their predecessor. Thus, we'll need to spend a long time to onboard and train the new employee. At the same time, there will be a lot of problems within the team and in relation to the client. Inevitably, the process of filling the competence gap will be painstaking. The costs are therefore much higher than a pay rise that could have been offered to the more experienced person that could have stayed in the team.

The problem is a complex one. You can't keep raising salaries of every person in the company. That said, it's crucial to give each case a lot of thought, evaluate the contribution of a given staff member and analyze the consequences of their quitting. Only then are you ready to make a decision!

The process cannot be generic and based on opinions in the vein of: "in my opinion, X is paid a lot of money already." This sort of approach is irresponsible and unfair. Each individual should earn as much as they're worth in this particular context and in this particular company. In order to get a sober view of the situation, it's a good idea to at least monitor the job offers that our people are getting on the market. And it's not about offering pay increases to everyone, yet it's worth considering whether our best people, key team members and future talents are adequately remunerated so that they don't even think about competitors' offers. It's often the case that even starting a conversation with another company causes a mental evolution that is impossible to stop.

In a nutshell, it's better to have a smaller team consisting of very experienced individuals (and consequently higher paid) than to manage a larger group of people lacking the necessary skills, who are still learning, but are theoretically cheaper. You have to be very careful with this sort of "savings."

Filling the Key Positions

In most cases, the best solution for an organization is to develop its own leaders. Owing to this approach, you can be sure that the person has a good understanding of the business and the organization they're in, has built relations with key employees, has earned respect and authority with their peers and people outside the team. Moreover, you can follow their development, watch their successes and failures, analyze their strengths and weaknesses, and even to some extent draw some conclusions about who they are as a person. In other words, the company has a lot of data on the potential leader (at any level) and the likelihood of making a mistake when promoting them to a key position is relatively low, even if not completely eliminated.

Promoting people at the organization is done by most companies. It works well if the company's growth is organic and stable, not too fast, when there's enough time for the natural leadership candidates to evolve at every management level. Problems start when the pace of growth is very high, all the best people have already been promoted and there are no more to choose from. Then there comes a time to start hiring experienced (and more expensive) people from the market. It's not only about leadership positions, such as team leader, project manager, or department head but also technical specialists such as senior developers, for example. If a company is growing exponentially, also technical competencies become more scarcely distributed. Experienced and stable teams are fragmented with the aim of becoming a core for new groups which will consist mostly of new hires.

In such a situation, it's a mistake to stick to old recruitment criteria which were good enough at some point, but aren't good anymore. While looking for a project manager for a large and demanding project with a tight schedule, for a new client, and the candidate would be expected to manage a large, recently formed team, they must be a competent, experienced person with the right skillset. It's extremely risky to fill this position with a random employee with a year-long experience of managing a team of four only because they're trustworthy and there's nobody else at the moment, and a specialist from the market would cost twice as much. We might save on one person's salary but then we'll be facing the chaos of a mismanaged project and that will cost us hundreds, if not thousands, in man-days. That's exactly why on such occasions, you need to be open to acquire experienced people from the market even if it disturbs the adopted compensation scheme, even if there's some doubt about how this decision will affect the organizational culture. Such people bring interesting experience from other companies and have interesting observations, they're more open to change, but of course they need to prove they can work at our organization and achieve the expected level of performance.

It's similar with development teams. If you don't want to spend money on filling the key positions, such as analysts, architects, senior developer, tech lead, quality engineer, you can't expect the team to be effective and their product to be of high quality. It's not about overpaying everybody. It's clear that no company is able to bear that. Yet key positions need to be filled with talented and really strong people. It's a good investment which will pay off as saved time for a lot of people. There will be fewer bugs, less fixing, fewer changes in the architecture, fewer misunderstanding in the agreements with the client, more assertiveness and confidence, etc.

To sum up, if the company is not able to shape leaders or generate and distribute knowledge in the technical domain, then you need to be open to hiring experts from the market, while accepting the unpredictable influence of this decision on the compensation policy and organizational culture. As what you get in return is some management, development, and innovation potential.

Summary

- A compensation policy tends to live a life of its own. It's often an impediment to taking logical and beneficial decisions, for example, it prevents hiring talented, competent, and experienced people, which also leads to losing such people in the long run.
- Projects are executed more effectively and less expensively with a smaller group of more experienced people.

- It's not a good idea to be exceedingly careful with spending when it comes to filing key technical, management positions or those which have direct contact with the client. The eventual losses are much higher than the anticipated savings, it leads to months in delays, additional man-days of work.
- It's not easy to estimate the value generated by an experienced employee, especially in terms of team and client relations. Before we let such an individual go for financial reasons, it must be carefully considered.
- The highest benefits and savings are generated by hiring talented, motivated, and competent individuals. Obviously, they are also the people who are highly valued on the market.

CHAPTER

22

When the Time Comes for Renegotiation or Terminating the Contract

The business side of any company starts and ends with hard-core analysis of its numbers.

—Bill Gates

At the onset of a project, we can usually anticipate two scenarios of how it's going to end: either we'll deliver everything according to plan, or the client will resign from our services before the project ends. The happy path does

M. Dąbrowski, *Managing IT Projects*,
https://doi.org/10.1007/978-1-4842-9243-3_22

not mean that everything will be done perfectly, meaning the scope will be delivered on time, within the budget, and with maintaining the expected quality, however the client prefers to define it. For quality is a very subjective matter. Yet, regardless of how well everything is going, in most cases we will encounter problems, generate delays, the scope will turn out to be larger than initially expected, users' expectations will be higher, business goals of the project might change in the meantime, the client might have to pay an amount several times higher than was estimated, etc. In short, it is relatively complicated to determine whether the project was successful or not. The client will take into consideration the quality of cooperation, the fact whether he can form a long-term partnership with the supplier, he will consider various alternatives in key moments and choose the one which will be optimal from a professional and economic perspective.

At the same time, the supplier must monitor the outcomes of the project. The first one that comes to mind is obviously the financial aspect, meaning whether the execution of the contract generates profits or losses. It's also a complex issue. Undoubtedly, it's best when the project is profitable. But what if it isn't? Then, it's important to consider some other issues such as the business development perspective and the chances that the client in question or some affiliated businesses have further buying potential. It's also important to assess the benefits of product development. It may turn out, for example, that the balance of the project is close to zero, but because we're delivering an innovative product which is at an early stage of development, in practice, the client is financing our R&D efforts, which means that in some way we're expanding our portfolio at the client's cost. It might happen that we're building a reputation in a lucrative region. So, even though we're not making a profit, the project is contributing to the development of the business by expanding our portfolio.

Contrary to what you might think, a negative scenario is not only the one in which the client terminates the contract with the supplier due to constant delays, the mistakes made, the fact that the supplier does not understand the business case, does not maintain high quality, etc. It is not often talked about, but it also happens that the supplier is the one who decides to stop the project for various reasons. If, for instance, the project generates losses, the client is unwilling to renegotiate the contract, is not responsive to our suggestions as to how to solve problems in a professional and honest way, and we can't see any other option, we can only try to end the cooperation in the least harmful way. Certainly, this sort of situation is never black and white, either. We need to take into account how the developed products and solutions affect our portfolio, the value of this particular client's testimonial, the impact the decision of terminating the contract will have on our prospects' opinion, etc. Nevertheless, the preceding scenario is always a possibility and it's good to remember that we have this option at our disposal.

When Continuing the Project Is Pointless

There are projects which turn out to be economically unviable after some time. It is seldom the result of the project team's poor performance, its members' incompetence, project manager's lack of skills, or some objective technical problems. Most often, it's the consequence of some serious mistakes made at the sales stage, wrong decisions made during the contractual phase, or inadequate risk assessment with regard to the region or the client's intentions.

A very common cause of losses generated during the project include the underestimated scope of work and wrong assumptions relating to the project execution as well as the lack of clearly defined financial mechanisms protecting the business. If the supplier takes on a challenge that he doesn't fully understand or does not have the necessary competencies and at the same time the client formulates his requirements very vaguely, the resulting contract is usually very general, it contains a high-level description of the scope, defined in such a way that each party can interpret the actual scope differently. Then, it turns out that the client's expectations were much higher, and what makes matters worse, these expectations are justified. The problem is that the supplier's costs increase significantly and he starts losing money. When the financial situation becomes too grave to handle, it's time to renegotiate the contract or to consider terminating it.

It may also happen that the supplier had miscalculated his own costs. The client's requirements were correctly identified, but carrying out the work turned out to be more complex than initially planned. Mistakes happen to everybody, even in business. When the client acts fair, is honest, and makes an effort to understand our problem and our context, a compromise might be reached, some additional revenue generated and the problem can be solved. Unfortunately, from a formal point of view, the client may decline to pay more than agreed. The client may also stick to the provisions of the contract. Then the supplier has no choice but to decide whether to continue the cooperation and if not, he should analyze all possible ways to finish the project and terminate the contract.

Another commonly encountered problem are clients who demand work outside the agreed scope. There are clients, specifically companies, for whom it's become a standard mode of operation. In some regions of the globe, it happens on a daily basis. Such clients constantly require extra features, argue about everything, demand more and more, they ignore substantive arguments based on the provisions of the contract. This leads to a substantial increase in the cost, delays, and changes in the invoicing schedule. Moreover, some companies withhold payments of correctly issued invoices as a way to get extra features for free. The supplier delivers features according to the schedule, formally approves the milestone, issues an invoice and then for a year or so asks for the settlement of the invoice, thus incurring extra costs, so that the client finally decides to pay. If we encounter a client like that, after some time it's worth considering whether further cooperation makes any sense.

There are also clients who – for reasons unknown to the supplier – don't pay the correctly issued invoices. If we deliver the scope and have formal proof for that, issue invoices and despite all that the client doesn't pay for several months, it's worth considering whether we still want to finance this cooperation. From an economic point of view, it's very risky. It might happen that the client has lost liquidity or the money has simply run out. To make matters worse, perhaps the client has already made a decision to stop the project but it's convenient for him to withhold this information from us. The client may just have bad intentions. Thus, if the client does not settle due payments, it might be a good idea to think about stopping the work, at least temporarily. The supplier can't continue the project if it jeopardizes the company's liquidity. Invoices are just documents which may turn out to be worthless pieces of paper. The client can, for example, terminate the contract, go to court, and question the issued invoices.

Another sign which may show that further cooperation is doubtful is the client's execution of various penalty mechanisms included in the contract. If penalties are charged for even the slightest violation of the contract, be it delays or not meeting the SLA, regardless of the supplier's objective efforts, the supplier should seriously consider the sense of continuing the work on the project and the attitude to the cooperation with this client.

Charging penalties is, first of all, a sign that the client doesn't care about maintaining good relations with the supplier!

The client's primary goal is to reduce the costs and he doesn't care about proving that it's different. He might even have prepared a long-term strategy for introducing a new player and replacing our products. That's why we should withhold any investments outside the agreed scope and rather try to reduce the costs. If the project generates only losses and we don't see any perspectives for getting a new deal which would cover them, we should face the facts and make a difficult decision about terminating the contract and stopping the work.

Nevertheless, before making any serious decisions, we should consider some of these issues. Does the project, despite the evident financial losses it generates, help us consolidate the market value of our business, the presence in the region, or give us a chance to get new contracts? Perhaps the client does not want to renegotiate the financial conditions, but still helps us to sell the implemented solutions, recommends and endorses our services? Life can often be surprising here.

It's also worth considering the impact of the unprofitable project on the company's portfolio. If, despite the losses, the developed product is innovative and strengthens the market position of our business, and at the same time, the client covers approximately 70% of the costs for developing the product, then in most cases, unless the liquidity is jeopardized, it would be a mistake to stop the project and terminate the contract. The project should be then treated as a subsidy to your business's R&D efforts.

One of the projects for a European company in the TMT industry turned out to be extremely underestimated because, as it happens, the scope was described in very general terms, and the client took advantage of that and persistently ignored any requests for renegotiating the financial terms. In theory, from the economic perspective, continuing the project made little sense. However, the project was about developing an innovative product, and interestingly the client supported the supplier's marketing efforts. Therefore, thanks to the seemingly unprofitable project several new lucrative contracts were signed.

Renegotiating the Contract

Negotiations whose aim is to enhance the contract terms are essentially similar to any other negotiations, but there's an important caveat. In this case, it's not an exchange of arguments, but rather making the client participate in them in the first place. Why would he do it? He already has a binding contract and the supplier should continue the work accordingly. So why would the client agree to worsen conditions? Why would he want to spend more money? In theory, it seems contrary to common sense, whereas in fact it's completely different.

Mature businesses develop supplier management practices over time. Managers understand that "starving" the suppliers or exploiting them won't bring any good in the long run. No business will proactively run an unprofitable project. It won't invest in cooperation without perspectives. In particular, the supplier will be moving competencies and the best people to other, profitable contracts. For the client, it will translate into minor reductions in costs as the amounts paid to this particular supplier will be lower, but in practice, it will cause delays, lower quality, less knowledge exchange, and continuous staff turnover on the partner's side. As a result, the client's losses will be much higher. His business goal won't be achieved on time and the delivered system might not meet the expectations, its users will work less effectively as it won't match their needs and its quality will be below par. Mature clients understand that the most rational strategy is to be fair toward suppliers. If they see that the project generates losses for the supplier, they will be open to discuss extra payments, within reason obviously. They understand that, in reality, it's in their best interest to let the supplier make profits. Only then will he be motivated enough to deliver.

The most problematic clients are those who avoid (or aren't able to for formal reasons) increasing the budget at all costs. Then the supplier should show that in an extreme situation, the project will be stopped regardless of what the contract says in this respect. Clients tend to forget that even if the contract does not stipulate any termination options for the supplier, there are still several different possibilities at his disposal. The supplier might, for instance, take advantage of any contractual provisions, formal agreements, plans, or dependencies and adjust the work in such a way that from a legal perspective the project will be frozen. Often, it's necessary to demonstrate what the cooperation could look like if there's no willingness to renegotiate the terms of the contract on the client's side.

When the client finally understands that it's a viable option for the supplier to stop the work, it paves the way for the exchange of arguments. The aim here is to show that continuing the project with the current supplier on new terms is still the cheapest and best solution for the client!

It's important to demonstrate that the process of choosing a new partner will be time-consuming. Firstly, it will only lead to delays in achieving the expected business goals. Secondly, one can never be sure that the new supplier will be reliable, what his products will be like, whether they will meet the client's expectations, etc. Thirdly, it must be assumed that the bidders will also want to make a profit, which means that it's rather unlikely that their bids will be more attractive than the current contract. And last but not least, the project will have to be set up from the beginning, started from scratch – the customer will have to pay for reaching the stage at which the current project already is, cover the extra costs and lose everything that has been achieved so far.

In most cases, it's possible to prove that continuing the project with the current supplier on new, improved terms is still the best and cheapest solution for the client!

If, however, new contract terms can't be negotiated, there's no hope for improving the situation, and further losses loom on the horizon, it's necessary to resort to the final step: stopping the project. In most cases, there's nothing else to lose, and the thing at stake is the business's financial standing.

One of the suppliers working for a European IoT solutions provider had been losing money due to the wrongly negotiated maintenance contract. The costs of work significantly exceeded the monthly revenue and, from a formal point of view, the supplier could not withdraw from the contract. After numerous attempts at improving the situation, many ineffective requests to the client for changing the financial conditions, the supplier informed that after the end of the current contract period, he will not prolong the cooperation and the maintenance service will be stopped. Because the contract was about to end in a few months the client realized that the alternative is to write a new tender for a similar system, select the best bidder, make an effort to implement the system which would probably be similar to the current one. Thus, it meant wasting a few years' work and costs much higher than continuing the cooperation with the current supplier on the conditions that the supplier expected. Therefore, the client finally accepted new contract terms.

Putting the Project on Hold

When the decision about stopping the project has been made, we have the following options:

1. Freezing the project indefinitely without terminating the contract
2. Ending the cooperation by mutual agreement
3. Stopping the project and going to court

Of course, we should aim to choose one of the first two options.

Freezing the Project Indefinitely Without Terminating the Contract

This option means that the work is stopped indefinitely so as to limit the costs and losses generated by the project. It is rational when the supplier is not able to terminate the contract (there are no such provisions in the contract) or there are no formal reasons to terminate the contract. It's a difficult option to implement. On the one hand, it's necessary to stop most of the internal work in order not to incur more costs. On the other hand, formally it's impossible to do and it's necessary to maintain some formal level of work for the client at the lowest possible cost. The supplier should take advantage of any provisions in the contract, the schedule and any stipulations relating to the cause and effect relations between the client's work and the supplier's obligations. It's some sort of a go-slow strike, all mistakes and violations on the client's side are pointed out to demonstrate that the work on the supplier's side is blocked.

It's important to remember that we're talking about a situation in which the supplier is losing money due to continuing the project for this particular client, and in which the client, despite being fully aware of the problem, ignores it and makes the supplier carry on the project anyway. It's a situation in which the client is not acting fair and the supplier has to defend themself. It's some sort of a test of strength. Most often the result of such a situation is either terminating the contract by the client (amicable or in court) or freezing the work by both sides indefinitely, which is often beneficial for both parties. Interestingly, it's not rare that the client returns to the supplier after some time and asks for resuming the cooperation while offering better, acceptable financial conditions. That's why this is a better option as we don't have to bear the costs and also the contract with the client is maintained, which gives us a chance to reach an agreement and resume the cooperation in the future. It is often the case, especially when there are some changes on the client's side (new management, new strategy, higher budget).

In one project after completion of the first phase, from the supplier's point of view, it became economically unreasonable to continue the project, carry out the consecutive two phases and lose more money. That's why there was a decision to stop the work indefinitely. However, the client seemed satisfied with the outcomes achieved so far, the system was in use and the client wanted to avoid conflict. Both parties maintained the status quo for several years. Eventually, the client's company was acquired by an industry giant, the managers on the client's side got much higher budgets for modifications in the IT infrastructure and the project was resumed on much better terms for the supplier.

Ending the Cooperation by Mutual Agreement

When both companies approach the issue maturely enough to accept and understand the other party's context, they know that further cooperation is pointless for objective reasons. Often clients themselves realize that their business case has become obsolete, or they have run out of budget for the project, they also know that they have no plausible arguments to go to court and end the project on better terms, avoid settling overdue payments or even demand a refund of those already made. For the supplier, termination of the contract by mutual agreement is beneficial on condition that some important issues are sorted out.

First of all, it's important to formally state that both parties decided to stop the project and it is not the supplier's fault. Secondly, it must be decided how the matter of costs of unfinished work shall be settled – the client should pay for the work as the supplier did the work in good faith. Thirdly, it's crucial to make sure that the client will not demand refunds of the already paid invoices. And finally, it's important to take care of the matters related to the Intellectual Property Rights, the deliverables (software, analytical documentation). It's not good when the client hires a new contractor who will use our analytical materials. This should be prevented at the stage of signing the contract.

All in all, business is business and the client, if he has a chance, will take advantage of any formal, legal, or business possibilities to finish the project on his own terms.

It's never the case that clients stop projects amicably! As was explained in the previous chapters, it only happens when the supplier has irrefutable, strong arguments and on this basis can present the project's history in his advantage. This means that the client decides to end the project by mutual agreement if it's clear that going to court will end in his defeat.

In order to ensure such a position, the supplier has to skillfully and meticulously gather the project documentation, make notes about the causes of delays, problems, decisions, especially in terms of the client's activity which led to problems on the supplier's side. It's obviously not about being vicious and doing things to spite the client. Absolutely not. Throughout the project, it's a good idea to be proactive, help the client, take ownership, but also to be aware of the fact that despite our efforts the client may find themself in a situation where they will stop the project anyway and sue the supplier. Regardless of whether we're acting fair, whether we make an effort, it may still end like that. That's why you have to be prepared from the start. Only strong arguments and the client being aware that we have them will ensure that the contract will be terminated on good terms.

Stopping the Project and a Court Trial

The matter of contract termination by the client was discussed in previous chapters. Here we will look at a case in which the supplier acts deliberately to stop the project, but the side effect of this approach is going to court.

It happens when the client seems to have enough arguments to get back a substantial part of the already invested money upon the termination of the contract. The decision to sue the supplier is usually driven by economic considerations. When the contract is terminated by mutual agreement, the client is obliged to settle all the issued invoices and also pay for any work in progress. Whereas by going to court, the client eliminates the need to cover the current supplier's costs. To make matters worse, the payments for any correctly issued invoices are withheld as the client claims that it's necessary to keep the money on account of the damages he will claim in court.

Besides, the client will surely try to get a refund of the costs incurred (licenses, invoices for implementation services), he will also try to make the supplier pay penalties, compensate for lost profits and the damages. All this is a game aimed at recovering as much investment costs as possible or even making a profit. In such a scenario, the only form of defense is gathering the project documentation which can help in building a strong case to support your position. Unless we're thoroughly prepared, we must be aware that the client will use the situation to his advantage. There's no mercy in business.

It is often the case that clients take advantage of their position as much as possible. Knowing that the supplier can't terminate the contract or stop the project, they threaten to sue suppliers and force them to continue the project and effectively keep losing money. It works until the sum of penalties and damages from the client is higher than the costs of continuing the work on the supplier's side. You must be incredibly smart, skillful at negotiating, and very lucky to defend yourself in this situation.

Generally speaking, if the supplier is losing money on the project and wants to stop it by terminating the contract, he must take into account that the client will go to court. Therefore it's worth being thoroughly prepared before making this decision: have strong and convincing argumentation, and then show the client the determination and readiness to fight. Shortly speaking, present how strong your evidence is. Only if the client sees that the supplier is prepared to fight and the client can actually be defeated, can we mitigate the risk of the court trial and make the client terminate the contract by mutual agreement.

All in all, if the losses generated as a result of continuing the work are higher than the potential damages that the client will claim, sadly we have no choice but to stop the work, even if it entails the risk of a court trial. The costs of the project are incurred here and now! The client, on the other hand, will have to wait for years before the trial ends. What is more, he will have to prove that his claims are valid, which is neither simple nor obvious.

Summary

- The supplier can also stop the project if it causes losses and does not offer business development potential either with this particular client or in general.
- Prior to making a decision about the termination of the contract, it's necessary to thoroughly analyze whether there are arguments for continuation of the work, it might be the case that thanks to the project we have a chance to finance the development of an innovative product or to get valuable testimonials which strengthen our brand and help us get new contracts in a particular region.
- The basis for renegotiating the financial terms is to convince the client that continuing the project with the current supplier is the cheapest and best solution, whereas the supplier is determined to terminate the contract regardless of the formal options and consequences.

- While preparing for stopping the project, it's important to try to reach an agreement with respect for both the client's business and particular people on the client's side. It is possible to freeze the project or to limit the scope of work, but it's a good idea to stay in the game and avoid terminating the contract. You never know what the future will bring. It might happen that the client will get a new budget, the management will change, some people on the client's side will change employers and invite us to take part in tenders written by the new companies.

CHAPTER

23

It's Always Good to Be in the Game

Success is not final. Failure is not fatal.

—Winston Churchill

Finally, some words about losing, giving up, surrendering, when we have no more arguments, possibilities, chances, and we need to come to terms with the reality. Unfortunately, there will be situations when the supplier is offended, destroyed, or blatantly cheated. Some clients will use unfair tricks, enforce extra work, make additional quality demands, deliberately prolong the approval process, blackmail, postpone signing the protocols despite all the work being done, indefinitely delay payments. There will be attempts at playing some unfair business games like enforcing discounts in fees guaranteed in the contract (either for the project or system support), attempts at optimizing the maintenance fee to lower the overall costs of using the system with an increasing number of users. Clients will introduce new subcontractors that will take over the maintenance and development of your system, they will

M. Dąbrowski, *Managing IT Projects*,
https://doi.org/10.1007/978-1-4842-9243-3_23

expect your support in implementing the strategy that is against your interest! They will insist on getting the source code to take control of the system and get rid of your company. They will demand services that will shorten the period of using your products. Your system will be slowly shut down and that will eventually significantly limit your business with the customer.

Confronted with this kind of treatment, people react emotionally at first, they get annoyed, engage in a conflict with the client, express their dissatisfaction, and thus ruin their relations even further. After that, they draw conclusions on the spur of the moment, analyze the current circumstances, wonder how the situation will affect the annual performance, the contract at hand, the anticipated revenue. The management tries to take advantage of the situation, force the client to pay additional fees, threatens, warns, resorts to blackmail and pressure, threatens to stop the project, uses financial penalties, for example, offers extremely high prices (much higher than reasonable), knowing that the client will have to commission additional services for the next six months either as a way of helping the new subcontractor or under the pretext of introducing the new subcontractor to the implemented products, integration services, etc. In this way, they try to make as much money as possible, as long as it's possible in given circumstances. The supplier sometimes resorts to some extreme measures and blocks the work for new players, thus also making life difficult for the client. It happens even that there are threats to sue the client.

All of these strategies, power plays, and desperate actions don't make much sense! They're based on emotions and a vision of short-term profits. They might increase the revenue for the next couple of months, but in the long run, they will probably cause irreparable damage to the relations and will effectively discourage the client from cooperation for many years to come.

After several years of conducting a huge transformation program with the goal of replacing an IT tool stack in a few countries of the group, toward the end of the project, the supplier started making losses on the projects. Power play was applied to force the client to pay more for the service. The client, having no other option, accepted the conditions. Initially, the actions brought a short-term positive effect, but the strategy was perceived by the client as arrogant, some people on the client's side felt offended and the relations of the companies deteriorated, they became colder, and finally both parties stopped their cooperation in mutual hostility.

In tough situations, you should try to act calmly, rationally, think logically, calculate what can be lost or gained in the long run. You should avoid thinking about lost dignity, hurt feelings, and injustice. You should consistently try to communicate and stay in the game! If it doesn't work today, you should make it work tomorrow! To make it possible, you have to be in the game.

The actions described earlier are not directly against you as a person or your company. You should try to understand the client, the situation at his organization, current strategy, act proactively, make constructive offers and at the same time clearly communicate that you as a supplier are losing revenue, losing business, and you can't afford to work for free.

Most often, the people who cooperated with you directly are happy with the project outcomes, you have a relation with them and they are the ones who make decisions about stopping the project, limiting cooperation, or offering less favorable financial conditions. You should always remember that! Most often, it's not their fault and if they could they would continue working with you, they wouldn't change anything.

The current situation is a matter of the client's financial condition, internal politics, particular people's mandate, certain managers' approach. Everything changes with time, it evolves. That's why you should talk as much as possible, negotiate constructively, be present at the client's company, go there often, have open communication channels and maintain at least some areas of cooperation. At the same time, you should consistently work on developing a new business opportunity, wait for changes and improved conditions on the client's side.

After completing a long and difficult project for a large company in the TMT industry, the supplier expected a period of stabilization and high profits. It turned out that the client demanded significant discounts in monthly payments. It was not elegant, not fair. After years of hard work and sacrifices for the project, when there was finally time to reap the fruits, there was an enforced decrease in profits. Obviously the client had his own hidden agenda – in the background, there was the transaction of selling the company to a new owner. As a result, the scope of the service changed. In theory, the supplier was treated unfairly, but he acted peacefully and constructively – worse financial conditions were finally accepted. After only a few months the supplier was invited to take part in a large project by the company which acquired the former client. Eventually, apart from not losing the current business he also gained a new reference and a prospect to grow. Whereas, if the supplier had acted aggressively, blocked lowering the fees (according to the contract he had a possibility to do that), he would have lost the client and any prospect of further cooperation with the new group would have been very unlikely.

When a client decides to replace your system, your product, you should also try to understand the reasons behind this decision. Maybe it's a business strategy on the capital group level, maybe there's no budget for continuing the work, or maybe your system is just worse, older, and less user-friendly. In such situations, you should act constructively, look for possibilities to stay in the game as long as possible. You shouldn't ever act in a hostile way, disturb, block, charge exorbitant amounts of money for your services to make it impossible for the client to carry out the plans – this approach will be a double-edged sword.

Of course, you shouldn't act naively either, cheer when the competition is taking away your business, or ask how you could help to be kicked out faster and easier! You simply need to communicate your position honestly and transparently, state it clearly that you're losing your revenue, explain why and what is not beneficial for you financially and business-wise. You should show which actions make absolutely no sense from your point of view and say what could be an incentive for you to cooperate. At the same time, you should avoid blocking the client and acting evidently against him (and in your own interest). You should look for a golden means, discuss, discuss, and discuss.

Theoretically, the client can always terminate the contract. It's only a matter of a smaller or larger mess on his side. Getting rid of the supplier from one day to another generates more problems, reducing the workload slowly and gradually is a smaller and manageable problem. If you act proactively, show loyalty and devotion, you explain to the client what causes you trouble, but you keep helping anyway the client will probably reciprocate, he will basically try to find a place for you in this new situation.

A client decided to replace the supplier's system. The supplier started blocking the client by charging exorbitant amounts for the services that the client had to order to switch from the current system to the new one. The supplier made a short-term income but the relations between the businesses were ruined. Only after several years and with totally new executives and managers on both sides was it possible to start over.

Life is full of surprises. What we take for granted today may be totally different tomorrow. While finishing or limiting the cooperation you should always bear that in mind. You should act in a way which shows that it's not the end for you, you can assume that one day you will work with your counterparts on the client's side again, if not in this company, it might be a different one, that it's not the last project you're doing together. You need to show your willingness to cooperate, your understanding for the situation in which they are, and help while being politely assertive. If the negotiations related to ceasing the project and terminating the contract is your last meeting how would you like to be remembered? You should take care of that.

In an IT project for a European company, a client all of sudden withdrew from the contract, right after positive results of the user acceptance tests of the first phase! It turned out that the client ran out of budget as it was necessary to invest more in another delayed project with a higher priority. There was an attempt to blame the supplier who had strong arguments and managed to refute the accusations. Eventually, the project was closed in a peaceful way, amicably, without a grudge, both parties declared being open for future cooperation. As it turned out, after almost a year, the client asked the supplier to submit an offer for resuming the stopped project.

I could give more examples to prove that it's not a good idea to fall out or give up, you shouldn't make impulsive or emotional moves which can only deteriorate the situation. Tomorrow will bring new perspectives and it would be unwise to ignore these opportunities due to hurt dignity or need to take revenge. As Winston Churchill said:

> *Success is not final. Failure is not fatal.*

Neither you're losing the business forever, nor is the competition in the winning position for eternity. That's why it's important to stay in the game!

Index

A

B

C

M. Dąbrowski, *Managing IT Projects*,
https://doi.org/10.1007/978-1-4842-9243-3

D

T

U, V

W, X, Y

Z

www.ingramcontent.com/pod-product-compliance
Lightning Source LLC
LaVergne TN
LVHW052311200926
848481LV00002B/526

9781484292426